MAKING OF A GENERAL

MAKING OF A GENERAL

A HIMALAYAN ECHO

Lt Gen **Konsam Himalay Singh**
PVSM, UYSM, AVSM, YSM

KONARK

Konark Publishers Pvt Ltd
New Delhi • Seattle

Konark Publishers Pvt Ltd
206, First Floor,
Peacock Lane, Shahpur Jat,
New Delhi - 110 049
Tel : +91-11-4105 5065
Mob : +91-93552 93900, +91-93552 94900
e-mail : india@konarkpublishers.com
website : www.konarkpublishers.com

Konark Publishers International
8615, 13th Ave SW,
Seattle, WA 98106
Phone : (415) 409-9988
e-mail : us@konarkpublishers.com

ISBN No.: 978-81-942018-8-5
Edited by Ranjana Narayan
Cover jacket by Idesign
Author's photograph ©Robin Wangkhem
Background images ©Kunal Verma
Printed and bound at Thomson Press (India) Ltd

I dedicate my memoir to ***Mangala*** *and our daughters,* ***Poornima*** *and* ***Priyanka****, who endured my long periods of absence during my 40 years of army life with fortitude, and a smile.*

To my ***parents*** *whose guidance and advice have helped me be the person I am today.*

To all the lionhearted soldiers of 27 ***Rajput*** *and all the formations that I had the privilege of commanding.*

And

To all the ***future leaders*** *of my great country.*

Contents

अजीत डोभाल, कीर्ति चक्र
राष्ट्रीय सुरक्षा सलाहकार
Ajit Doval, KC
National Security Adviser
Tel : 23019227

सत्यमेव जयते

प्रधान मंत्री कार्यालय
नई दिल्ली–110 011
PRIME MINISTER'S OFFICE
NEW DELHI - 110 011

September 28, 2020

MESSAGE

Gen. K. Himalay Singh's book celebrates the extraordinary journey of a small boy from a sleepy hamlet in Manipur to become the first Lieutenant General of the Indian Army of North East India. Replete with sage wisdom and tales of grit, I am confident the work would inspire coming generations of our army leaders to attain high professional standards and dedicate themselves to the service of the nation.

(Ajit Doval)

Pre-Publication Praise

The book is an elegant narrative of the transformative years of a young boy from Manipur and his rise to the pinnacle; as a Three Star General. Leading his men along the Northern borders as well as in Counterterrorism Operations, Konsam has been a Gentleman General, with the ferocity of a 'Tiger'. (In doing so), he has chronicled his views on Ethical Leadership with an underlying message to 'Carry out one's duty and face challenges Head-on'. I am sure that this success story will be a motivating read for the youth of India.
—Gen. Bipin Rawat,
Chief of Defence Staff

Making of a General is a passionate and deeply moving memoir, cataloguing the journey of a village boy from Manipur who dreamed of serving his country and did so with utmost distinction. Writing the story of one's life is a daunting task, yet Lt Gen. (Dr) K.H. Singh (Retd) managed this arduous task with remarkable frankness and alacrity. The narrative of service rendered during the Kargil War and in the Siachen Glacier and as a Corps Commander in the LoC (Pakistan) is an account of perseverance and excellence and should be a source of inspiration to all, in particular to the youth of North East.

The book is sure to earn the unqualified admiration of anyone who is curious to learn more about the character and depths of the Indian Army and the proud men and women who serve in it.
—Mr Ranjan Gogoi
Former Chief Justice of India

Do hope that the book reflecting the author's military experience of four decades, zeal for learning and academic distinction and his work in civilian sphere (in Manipur) shall be read by all those who wish to serve the nation.
—Mr N.N. Vohra
Former Governor of Jammu and Kashmir

It is a tale of perseverance and grit—from a 'physically weak, shy boy who was scared of heights and darkness', to rendering his services in the Kargil War and in the Siachen Glacier. The chronicles of his career, replete with challenges, pursuit of knowledge and understanding, his astute observations of the geo-political developments in India's neighbourhood, his pondering over the issues that plague his beloved Manipur—a place of so much promise with thwarted dreams—is a call to duty for the younger generation to take up leadership and to actively engage with problems and attempt to find lasting solutions.
—Ms Pooja Elangbam, IAS
Youngest and Third lady IAS officer from Manipur
Founder & President, Book Club Imphal

This book should be an inspiration not only to a generation of aspirants for profession of arms, but also to all those who are fortunate to serve in the Indian Army, an institution beyond compare.
—Mr Maroof Raza
Defence Analyst

Introduction

One day I will be a Brigadier are the words I wrote in my diary in 1967 dated 26th January—perhaps the first chronicled entry of my life. I was then studying in an idyllically located school in a remote village of Charangpat (Lake of the Dragonflies) in the south eastern foothills of Manipur Valley. I was physically weak, shy, and scared of heights and the darkness. Many a mate in my school mocked my physique and ambition of wanting to join the army, leave alone rise to be a Brigadier. Thus, it became a relentless pursuit of self-improvement throughout my formative years in the school days. My yearning for physical fitness, academic and intellectual excellence persisted through most of my early life. Fifty years later, I returned to my roots in Manipur after having served in the Indian Army for nearly 40 years and after having become the first Lieutenant General of North East India.

As a young Lieutenant commissioned into the elite Rajput regiment of the Indian Army it had always been my dream to jot down my life's journey. As a subaltern, in my quest for adventure and also to get over the fear of heights, I ventured on many mountaineering expeditions, trekking in the Himalayas including an expedition to Mount Kanchenjunga, and skydiving trips in Agra. Later, I led an Infantry Battalion during the Kargil War and commanded a Battalion in the highest battlefield of the world, the Siachen Glacier. The memories of the two years operations in Siachen and in Shyok valley with my troops are truly the finest hours and unforgettable days in my life as a soldier. Still later, I commanded an Infantry Division and a Corps in the Line of Control facing Pakistan and also oversaw Counter Terrorist Operations in Jammu and Kashmir.

My professional exposure also included sensitive responsibilities in the South Block and Sena Bhawan in Delhi where I witnessed policymaking at higher levels from close quarters. There were times when matters did not navigate as one wanted them to. In any event, I felt I contributed my penny's worth.

My interaction with the cadets as Battalion Commander in the National Defence Academy, Khadakwasla, and later with the officers as Commandant of the Infantry School, Mhow, gave me important insights about leadership training and professional development of future leaders of our army. During my tenure in the higher ranks I became increasingly mindful about the effects of the existing personnel policy in the army and its impact on national security.

This book aims to impart some thoughts and lessons that I was able to garner for the future generations. The issue of 'Ethical Leadership' development in the army has been analysed in some detail. Lessons learnt from routine interaction with the soldiers turned out to be change catalysts in my man-management style that I adopted during my various command years. No less were the learning experiences from my colleagues and seniors. The book also attempts to highlight the challenges of the army from across the LoC in the areas South of the Pir Panjal Ranges in Jammu and Kashmir during the period as well as in the hinterland.

Indeed, what does it take to be a General in the Indian Army? Does it require one to excel in every possible activity in the organisation? Is it possible for anyone to reach the exalted status with hard work and dedication alone? Are the senior officers adequately equipped with the personal and strategic leadership qualities that are expected of them for the onerous responsibilities that lie ahead? Some of these issues have been covered in the backdrop of my direct experiences in dealing with such subjects while I was in service. I have attempted to pass on my experience in the higher echelons of the army and my views on Generalship in order to enable the future generations to view the horizon in bolder light and take informed decisions when time beckons. The recent geopolitical developments in the region and the aggressive manipulations of China and Pakistan have also been touched upon in the backdrop of my direct experiences in dealing with some of these issues while I was in harness. I am, however, not claiming to have analysed in detail the contours of the strategic issues in our neighbourhood.

I have also attempted to analyse the positive, personal attributes of the youth of Manipur that has made the state secure for itself one of the highest per capita number of officers in the armed forces of India in the present time.

My return to Manipur after superannuation also fulfilled the promise I had made to the late Dr A.P.J. Abdul Kalam, former President of India. The undertaking was made when my formation in Delhi had the honour of hosting him for a few months until the completion of the repair works of the house allotted to him in Lutyen's Delhi after he left Rashtrapati Bhawan. Dr Kalam had discussed and educated me about the political and socio-economic future of Manipur. He—at the time—did not have great hopes from the contemporary polity and the ecosystem in the Manipuri society of that time. He, in fact, felt sorry for the people of Manipur. My return to my home state and my experiences in a constitutional assignment as Chairperson, Manipur Public Service Commission and in the state's Consultative Group on Naga Talks confirmed the fears that the late President had expressed to me in person way back in 2007. It was my desire to reflect on my earlier exposure to the insurgency-related problems in Manipur and Mizoram and to view the problems through the prism of yesteryears. My research for the PhD thesis on *Ethnic Conflict in Manipur* also provided me the necessary academic content for the personal perceptions and experiences as a soldier to face the myriad issues of Manipur. Having been a soldier throughout my life, I took up some of the issues in my sphere of influence head-on. Some of my objectives were achieved. Unfortunately, some have been such that there has not been much to celebrate about.

I felt the more I began to understand Manipuri society and its inherent foundations, the more I realised that we were slipping away from the path of progress. This was largely due to the emerging contradictions in the society and dreams of the populace that made up the present generation. The aspirations of the people are greatly removed from the ground realities that I have witnessed in recent times. As I have preambled, it had always been a dream for me to pen my thoughts on the present and future trajectory of Manipur and the North East.

The book also attempts to highlight and portray the changes taking place in the socio-economic milieu at an amazing speed from the period of my childhood to date. It captures the effects of environmental degradation within a very short span of time. The state of instability and conflicts of the past few centuries which have shaped the emergence of what Manipur is today have also found resonance in my jottings. Many a well-wisher encouraged me to share my varied experiences in the army and in the civilian street with the future leaders of the region. My wife and my daughters, not used to seeing me with plenty of spare time, encouraged me to share experiences of my childhood, my adventures and military journey in the Himalayas including those during the Kargil war and, of course, the making of the first Three Star General from India's North East.

The purpose of the book has also been to portray the 'Land of Jewels' of yore and my salad years in the land of my birth. My growing up years in a joint family and learning the life skills in the natural settings of the land have been covered in some detail. My sojourns through the nooks and crannies of the hills and mountains in the Higher Himalayas as also in the hills and woodlands of Mizoram, Manipur, Sikkim and Kashmir where I spent most of my days in the service of India find an important place in the book. The chapter on Kargil War as I saw it and the two years in the 'highest and the coldest battlefield of the world' is a tale dedicated to all those soldiers who I had the privilege to command during war.

The memoir is also about the life of a soldier replete with joyful songs, be it in the Line of Control or in war/warlike conditions, believing always in *Nishkaam Karma*. My endeavour has been to reach out to the youth and say that no matter what you do, success ultimately is about carrying out one's duty and facing the challenges that come in life. It is also to arouse the great potential among the younger generations in the North East so that they may participate in the great Indian journey. The Postscript along with Jaideep Saikia, a renowned author and security analyst based out of Guwahati, Assam, is aimed towards such a message.

The views expressed in the book are mine alone. The love and affection showered on me by the people of my state hugely influenced my desire to write. I would have achieved my objective if, after reading my book, even a few young minds are inspired to face challenges in their lives and in society with a smile and realise their dreams of making an evolved society in the future. A song can change a life, a flower can spread the fragrance of life, and a good deed can inspire another life. The work is aimed to convey a message with a deep sense of conviction—that there is tremendous potential among the young people in the North East.

I am conscious of the fact that my small contribution to society through the medium of this book will at best be a drop in the ocean. Yet, I do believe what Mother Teresa said once, 'We ourselves feel that what we are doing is just a drop in the ocean. But the ocean would be less because of that missing drop.'

The book is dedicated to all the valiant colleagues and soldiers with whom I had the privilege of being with in very challenging times. And to Mangala, Poornima and Priyanka who withstood many storms with courage during my long absences on duty away from them.

1

The Foundation

Takane No Hana

Deep in the luxuriant blue hills and green valleys in the junction of India and Southeast Asia where the Siroy Lily blooms and the Sangai (brow-antlered deer) dance is a land called Manipur. Lord Irwin (1926-31), the then Viceroy of India, described Manipur as 'Switzerland of India'. The Japanese soldiers who fought in Manipur during the Second World War described it as ***Takane no Hana*** (Flower on the Lofty Heights). Manipuris were called *Kathe* by the Burmese, Cacharis called them *Moglie*, while the Ahoms of present-day Assam referred to them as *Mekle*. The Shans of Burma called Manipur as *Cassay*. There were many names associated with present-day Manipur—*Meitrabak*, *Poirei Meitei leibak*, *Kangleipak*, *Mekhla* and so on. The hill ranges of Manipur form part of the Sub-Himalayan Mountain system. Though landlocked, Manipur is believed to have been the route through which the South West Silk route between ancient Chinese and Indian civilisations have traded via northern Burma.

Charangpat Village (Lake of the Dragonflies)

I was born in the remote village of Charangpat (Lake of the Dragonflies) in the southeastern foothills of Imphal valley. I am told that my great-aunt named me 'Himalay', a very uncommon name in society at that time. My grandmother told me that naming a child after the mountains brings good luck. There is a beautiful hill range a few kilometres away east of our house. Maring and Tangkhul tribes are the dwellers of these hills. Mountains and hills have inspired me even as a child. Water bodies and paddy fields surrounded the village in abundance. Arong Turel (river) runs through the length of the village. The river is named after the local deity *Arong Ningthou* (King of Arong), one of the many *Umang Lais* (forest deity) in the area.

I recall the beautiful memories of my childhood in the village. Charangpat is a small village situated 25 km from capital Imphal. It was inhabited by a little less than 1,000 people in the year 1957 when I was born. There was a high school in the village. Literacy was very low. There was no doctor, no engineer, no electricity and water supply, and no post office. The village was connected by a *kutcha* (mud) road to Thoubal, 4 km away, from where regular bus service connected to Imphal town. There was a lone bus at six in the morning which ferried passengers going to Imphal and returned in the evening. The bus belonged to our family for a few years. The village was well endowed with natural drainage water system from the small rivulet and from the paddy fields during the rainy season. People were happy with what they could get from the paddy fields and from their respective *ingkhol* (homestead land). The *ingkhols* were self-sufficient in terms of a pond with lots of fish and vegetables for consumption of the whole family. Bamboo and many trees surrounded the *ingkhols* as some kind of fencing. *Upan-wapan* (natural screens of trees and bamboo) was all around the homestead land. Migratory birds like egrets and storks adorned and covered the entire *Upan-wapan* during their migration. We watched the migratory birds with delight. Alas! These scenes would appear to be fictional and romantic today. Cows, bulls and Manipuri polo ponies roamed the place too. We played in the big courtyard. My grandparents would watch their grandchildren play in the rains here. In the nooks and corners of the house and out-houses, there were many items of military use since the fields nearby were once used by the Allied forces as airstrips during World War II. Remnants of parts of fighter planes of the World War days were seen in the small culvert connecting our house and my uncle's house over a drain. I saw some rifles and ammunition rounds which were remnants of the war kept in the *Yumjao* (Big House). All these warlike stores were deposited with the police during my childhood days. My grandmother called me one day to show me her bundle of notes which she had hidden for over 10 years from others. The bundle was of 100-rupee notes printed by the Japanese government as currency for the Indian National Army (INA) during the war. These notes were not in use after the Great War. In fact, some of them were given to some people to keep them in the museum. In our paddy fields too, we could find remains of many small parts of army vehicles. Agriculture was the occupation of most of the inhabitants of the village. Very few were in government service.

My grandfather was a prominent personality there. His word meant a lot to the needy people in the village. He was frequently invited to various social and religious functions. I accompanied him a few times to these events, and saw the reverence the people had for him. Even the horses kept silent when he coughed. I recall that he contested the legislative assembly elections sometime in 1964-65 without success. He ensured that any Maring or Tangkhul tribesmen passing through the village were well looked after in our house. We were required to provide water and bedding arrangements to them in our courtyard while transiting through the village. Ours was a joint family. My father being the youngest son, it was customary to reside with my grandfather. I being the first child of the youngest son of my grandparents, I think they treated me with maximum care. We are five brothers and one sister. My third brother Khogen passed away in 1978 during his studies in Lawrence School at Sanawar, Shimla. He was a brilliant student. He had secured a scholarship to study in the United Kingdom after Class 11.

My mother used to take us to Poirou Khongjin village near Kaina hills where my maternal grandfather resided. It involved about 2-3 hours of walking from our house to the village. The crossing of River Imphal en route was on a small hand-paddled boat. My maternal grandfather often took me for fishing in his part of *Poirou Pat* (Lake Poirou) nearby. In fact, fish was abundantly available in the lake. We had to just put a fish basket inside the water body and take a fully-loaded basket of fish back home. He also took me to some of his Tangkhul friends and showed me their pineapple farms and pig rearing too. My great maternal grandfather built and dedicated the first temple at Kaina hills. It is said that Lord Krishna appeared in the dream of the Maharaja Bheigyachandra (King of Manipur during the 18th century) and the Lord is believed to have conveyed to the king that he would be in the form of a jackfruit tree at Kaina village. The Lord commanded him to make his idol (*murti*) from the same tree at Kaina. The king accordingly got the *murti* made and the same had been installed in the Govindaji temple at Imphal. Since then the management of the Kaina temple board was entrusted to the Pangabam clan to which my maternal grandfather belonged. On my further inquiry about the association of the *Sagei* (clan) of my mother with Kaina, I learnt from Uncle Babu (90 years old, retired secretary of the Manipur Legislative Assembly) that my mother's clan used to inhabit Kaina area and looked after the jackfruit tree till the legend of Lord Krishna and Maharaja Bheigyachandra got associated with Kaina. My mother's clan then shifted out from the location and settled in Poirou Khongjin a few kilometres away. Uncle Babu also told me that the jackfruit tree from which the deity was carved actually grew from a small sapling offered to the *Phunga-Lairou* (fireplace inside the house) by my great-great maternal grandfather inside their house at Kaina. The same sapling grew to be a big tree many years later. I found in recent years that a lot had changed in the area. Many efforts had been put to further develop the area.

A Calcutta-published newspaper *Anandabazar Patrika* subscribed by my eldest uncle the late Shri Gulab reached our place 2-3 days late. He was the Headmaster in the Higher Secondary School at Thoubal. With little knowledge of the English language, I learnt about the many events around the world through the various photographs of the events displayed in the newspaper. My uncle took pains to explain to me the English news. Many events in India and the world amazed me during my early childhood. The Apollo 11 had landed on the moon and the Vietnam War was in full force by the Americans. I saw grim pictures of the fighting in Vietnam in the newspaper. Thousands of soldiers and civilians perished in the Vietnam War. India was in the midst of a large-scale famine including in my tiny state of Manipur. The Indo-China war of 1962 had a significant impact in the minds of young people like me. I heard stories about the Chinese intent to snatch away the northeastern part of India. I came across one local soldier who had returned from the Chinese war. He narrated how the Chinese soldiers attacked our forces in the Himalayas. In my state, struggle for what many called independence of Manipur was gaining momentum in the late 1950s. The insurgency in Naga areas of Assam was at its peak. We heard many stories about the fighting between the Indian Army and the Naga Hostiles mostly in the hill areas of what is now Nagaland. Many army soldiers had been killed. Many hostiles (the term Naga Hostile was coined by the army) also perished in the fighting. The fighting had spilled over to the northern parts of Manipur Hills as well by 1964, the time when I was able to understand things. The hills which were adjacent to our area were also known to have many hostiles hiding or operating in the vicinity. We also heard rumours that Chinese forces were likely to enter our area to help the Naga groups who were fighting the Indian Army.

The Government of India raised a force named Village Volunteer Force (VVF) during the time. By 1967-68, the VVF under one R.K. Madhurjit Singh of Imphal was organising arms training camps near our village. Most of the young children above 12 in the village were taught the use of Bolt Action 303 Rifles. In the Fifth Standard, I got the first opportunity to use and fire a gun near the village of Langmeithet which was located in the foothills. I presume that it was an effort by the government to thwart the activities of Naga Hostiles or even to resist possible Chinese invasion! I was only 10 years old studying in a village government school in Charangpat. I wonder if I was eligible to handle weapons at that age under the law. Many of my friends in school also took part in the arms training.

In 1964, after the China war my uncle who was an engineer in NEFA (North-East Frontier Agency, present Arunachal Pradesh) came to visit his father, and our grandfather. He called me and narrated some stories of the war as he had seen there. He was serving as an engineer at a place called Roing in NEFA. He used to encourage me to do well in studies. He asked me what my aim in life was. It was a tough question since I had not given the subject a thought at that point in time. Nonetheless, I replied that I wanted to be a famous personality like Mahatma Gandhi! He was obviously very amused. He advised me to pen down my aim in life. He told me that I should reach the pinnacle in the field that I chose. He told me that I should aim for a

Nobel Prize if I wish to pursue a career in academics; participate in the Olympics if I wanted to be a sportsman; and become a Brigadier in the army if I chose to be a soldier! He advised me to buy a diary where I could pen down my thoughts. After procuring a personal diary that year from PC Jain stores in Imphal, I made my first entry in that diary.... ***One Day I Shall Become a Brigadier in the Army***. The entry was dated 26 January 1967. The idea of soldiering was very close to my heart. The rank of Brigadier was very prominent in the entire North Eastern Region because Col Sailo of (Lushai Hills) Mizoram had become the first Brigadier in the army from North East India around the time. I was told that many among the Naga Hostiles wanted to become self-styled brigadiers in the outfit. Brigadier Sailo's elevation to the rank of a Brigadier in the Indian Army made news all over the North Eastern Region, hence my desire increased manifold to become one as well. Brigadier Sailo later became the Chief Minister of Mizoram.

Ours was a joint family. My uncle's family comprised three sons and two daughters. So in all we were 12 members when I was 10 years old. Our home was a modest *Yumjao* (large village house). There was no electricity in the village. In fact, some of my cousins went to see the first domestic electrical connection in the area at Athokpam, a neighbouring village near Thoubal town. We were all excited when we saw my cousin Bhorot press the switch and light came! There were a few lanterns and simple *podons* (bottles with wicks) in the house. There were a few polo ponies in the stable, a few cows and bulls were there too. It was like a farmhouse or a ranch. We were encouraged to ride horses and play *Sagol Kangjei* (traditional Manipuri polo) by our grandfather. The best behaved horse was a white one which our grandfather was very proud of. We were taught to look after the horses and cows in particular. The homestead was self-sufficient with sugarcane, lots of vegetables, two fish ponds and a few fruit trees.

I remember going to the nearby paddy fields to see our parents work in the fields, particularly during harvest season. The paddy fields were very fertile. Plenty of fish used to bc found in the paddy fields during the latter part of the monsoons. We learnt to catch fish from the fields and from the irrigation drains with the use of *Loo* (bamboo basket traps). We were expected to collect the dried cow dung from the fields for use as firewood at home. I used to follow the elders in the group for the same work. There would have been hardly any impact on our livelihood even during a prolonged lockdown! Every family home had adequate means to survive in such situations. At home, we also used to make sticks of cow dung for use as firewood. On average we used to make 15-20 sticks each in the morning before proceeding to school. On return from school we were required to water the vegetable plants in the *Ingkhol* (kitchen garden). Thereafter, it was customary to put the lighted lantern near the *tulsi* (holy basil) in the middle of the courtyard just at sunset.

Vaishnavite Flavour

My grandmother was a very devout Vaishnavite Hindu who never permitted us to cook or eat non-vegetarian food except fish inside the house. However, we were allowed to cook and eat chicken dishes outside the house. Seeing the changing habits and preferences of the young people like us, she started giving permission to boil eggs inside the house. She even allowed me to own a small chick which I picked up from the roadside one day while returning from school. She justified her decision by saying that her grandson (myself) is doing a great thing by saving the life of a chick! During *Yaoshang* (Holi), it was customary to build a bamboo thatch hut in the locality for the young to congregate and hear stories (*Funga Waris*) from the elders. These bamboo huts were made by stealing the bamboo and the thatch from nearby localities. It was also customary to gather vegetables at night time from nearby houses without the permission of the owner. More like stealing these vegetables and have mini eating parties! The thrill of stealing during the festive season was at times short lived by some owners catching the intruders and getting beaten up. Perhaps the tradition of stealing the things during the *Yaoshang* (Holi) festival could be linked and justified to the stealing of butter by Lord Krishna in his childhood. We used to play battle scenes by creating two warring sides armed with fireballs. These fireballs were nothing but the dried ant colonies found in the trees. Frequent sports activities in the localities were regular, and I am happy to see the sports activities being perpetuated till now. The tradition of making bamboo thatched huts during Holi is slowly disappearing. Most people hardly know the real significance of Holi festival in the village. This is because Hinduism in Vaishnavite form was introduced in the kingdom only during the 18th century by King Meidingu Charairongba. It was taken as will of the king to embrace the religion. Two Brahmin families in our village were responsible for all the religious rituals in the village.

The Brahmins recited the verses either in Sanskrit or Bengali during the religious ceremonies. Only a handful of the residents of Manipur understood these languages. But the *bhakti* of the spiritual platform provided them the required spiritual needs and thus continued the practice of Bengali and Sanskrit verses for many years. However, with the changing times I find that the verses in use nowadays are mostly in Manipuri language. Hence, it can be assumed that most festivals were celebrated in the past as *Kumei* (entertainment activity). Most of the Hindu festivals were celebrated in some locally modified form or the other. Many local deities also had corresponding names of Hindu deities. *Baruni Ching* or *Nongmai Ching* (a revered hillock in the eastern valley area) was also Lord Mahadev's abode. Kondum near Sugnu in the southern tip of the valley was also pilgrimage for worship of Lord Mahadev. Such was the belief and practice of Vaishnavite

Hindu culture among the masses that common people were mostly ignorant of the Hindu religion itself but practiced the same as rituals and as tradition of the Meitei people who followed Hinduism. I played the role of Lord Krishna at age 11 in the *Ras Lila* (a five-day dance festival in praise of Lord Krishna) in the village. It was considered a privilege for any boy or girl to play the role of Krishna or Radha in the *Ras Lila*. I recollect that I had to memorise the *slokas* (hymns) in Sanskrit without understanding even an iota of the meaning during the five-day dance and music activity. I still wonder how I managed to memorise the Sanskrit *slokas* which I delivered flawlessly as I danced as Lord Krishna during the *Ras Lila*.

Lessons from Childhood

My first school was the Charangpat Junior School located just a km from our house. I recall that the qualifications of many of the teachers in the school were 'Appeared'. I wondered what this meant. I was to learn later that they had all 'appeared' for the matric examination but failed! I was the youngest in the class. But I topped in academics. Comparatively younger and weakly built, my inability to win in games and sports made me deficient in self-confidence. The *Yaoshang* sports winners went home with attractive prizes like a lantern or a vest. I always wanted to buy a lantern for my mother. Whenever my mother took me to her relatives in Poirou-Khongjin and Yairipok, I was given a few coins as reward for my excellence in studies. I collected all these. One day I mulled as to what to buy for my mother with a princely sum of 75 paise that I had collected over a few months. I went straight to a local vendor and ordered salt for the entire amount. The shopkeeper asked me if I really needed to buy salt for the entire sum of 75 paise. I said yes. I brought with me a huge packet of salt for my mother. I could not afford the *Meitei Thum* (locally made salt) since it was more expensive. She was astonished and asked me why I bought such a large quantity of salt. I told her that I had seen grandmother telling her to use salt sparingly, so I made the purchase to save her from grandmother's comments. This incident became the talk of the village elders!

I made an endeavour to shine in other fields too besides being a topper in class. I took part in a poetry recitation competition at age 8 or 9 at the state level. The title of the poem was *Khongjom Turel Yaiphabi* (The Blessed Khongjom River). The theme of the poem was that River Khongjom was blessed by being washed by the blood of the martyrs of Manipuri forces who laid down their lives fighting the British in the Battle of Khongjom 1891. The huge audience which I saw as I commenced my recitation gave me the jitters. Fear of the audience overcame me and I started crying as I delivered the recitation! Lo and behold, the crowd was spellbound by my emotive delivery! My tears rolled perfectly in harmony with the theme of the poetry and I was adjudged second in the competition. I was elated but I knew I didn't deserve it. I simply cried due to fear of the audience while reciting the poetry. I knew I had to work far more to improve myself in every aspect. I recall another incident of attempt to cheat during an examination in the Fourth Standard! The older boys told me that they were not bothered about the examination because they could easily pass by copying and cheating. I was confident of topping in the class. But I wanted to experience the thrill of being part of the larger group that was dependent on cheating in the examination. So I carried a small chit with notes during one of the examinations. I became very uncomfortable with myself inside the examination hall and wanted to throw away the chit inside my pocket and asked the teacher to permit me to go outside. He didn't suspect me since I was good in academics. I threw the chit away and was relieved. I continue to reflect on this particular incident as to what was my motive in the whole episode. Did I act owing to moral considerations? Did I act due to fear of getting caught? I think it was both. I learnt my lesson in self-improvement early in life. The line between acting out of 'fear' and due to 'moral considerations' appeared thin.

We were required to write the school board examination at the state level in the Fifth Standard. During that time Manipur was a Union Territory. It gained statehood only on 21 January 1972 after a long struggle by the people of the state. I secured ninth position at the state level in the Fifth Standard examination. My position earned me a state stipend, a princely amount of Rs 5 per month. It was a considerable amount for me that time. In fact, the same year I was the first to wear a pair of shoes in the village school. My mother bought me a small plastic made globe of the planet earth. I was very fond of searching for locations in the globe! Playing with the globe gave me immense knowledge about the geographical information of planet earth.

During the great famine in Manipur during 1965, the situation was bad. All Manipur Students Union led the agitation against the authorities to stop export of rice. Four students died in the agitation at Imphal. The demand for statehood was also gaining much ground among the young in particular. My cousin brothers, Baru and Michael, were very actively involved in the agitation for statehood demand. Nagaland had been given statehood earlier in 1963. The people of Manipur felt humiliated that a 2,000-year-old kingdom which was merged to the Indian Union in 1949 was denied statehood for a long time. The much-awaited statehood was granted later in 1972.

My uncle the late Sri Apabi was a famous polo player of Manipur in those days. I had not seen grandfather playing the game though I knew that he was a great player. Uncle was awarded Rs 3,000 annually as a token for his contribution to the game of polo from the state government. It is a well-known fact that the game of polo originated in Manipur. Iran and Mongolia apparently also claim that the game originated in their respective countries. Be that as it may, I wish that our own country does more to preserve the original game as played by the Manipuris. The International Polo rules are quite different from the rules that were followed in the traditional polo in Manipur. The Manipuri Ponies are becoming near extinct today even though the authorities are making some efforts to preserve the breed as well as the game itself. Too little, too late. I learnt that there are only a few elders left now who know the traditional game well. The Manipuri traditional polo does not enjoy the rightful place, as most polo tournaments even today are played as per the international game of polo, even at Imphal. Today, I am happy that All Manipur Polo Association, despite its very meagre resources, has taken upon itself to revive the indigenous game of *Sagol Kangjei* (Manipuri polo).

When I went to the Sixth Standard, I was admitted to the Chaoyaima Higher Secondary School at Thoubal, located on a small hillock some 4 km away. The school had better facilities than my village school. Teachers were more qualified, and my eldest uncle Shri Gulab was the Headmaster. My uncle ensured that strict discipline was maintained in the school. Latecomers were punished. He also ensured that all the students were properly dressed. Absentees were minimum. The class was more competitive than my old school in Charangpat. My cousin Roben (now an accomplished Merchant Navy officer) also took admission in the same class. We walked for 8 km both ways every day to and from the school. The 4-km-walk to school every day through the lush green paddy fields was a joy for me. At times thunderstorms, strong winds made our walk uncomfortable. My friend Munal (he retired as Subedar from the army) had a knack for locating snakes on the way while crossing the paddy fields and he was extremely good at catching them. At times we got to chase and jump on board the slow running bus to stand on the rear bumper en route to Thoubal. The bus driver permitted us to do so on the condition that we regularly filled up the potholes in the 4-km-long kutcha road. We were not, however, permitted to board the bus and sit inside. I made many friends at school. Teachers were very hard task masters.

I recall an incident which reflected my inner conscience on one side and the will to be the numero uno in every subject. I was unhappy that I secured 59.5 marks only out of 100 in one of the English examinations. I was determined to cross the first division score of 60 on my own steam rather than get a grace of 0.5 marks from the teacher which anyway would have been done by the teacher. I went over to the teacher, the late Oja Sanajaoba, to claim that my answer to one of the questions deserved 0.5 marks more. He understood my determination to get the 0.5 marks. He counselled me that I deserved 59.5 marks only and that 0.5 marks more to make my score a figure of 60 was the prerogative of the teacher. He further advised me not to ever beg for anything extra from anyone in life. The same value was imbibed and reinforced by my parents. I maintained my position in the class by claiming the top spot in most of the tests. The spirit of competitiveness to excel in the class became more and more intense in me.

The first local club in the village SEOB (Students Endowment Organisation Board) was conceptualized by my elder cousins Ta Sarat, Ta Baru, Oja Kalidas and a few others around 1965-66. I was one of the boys who was given the task of hand-copying selected books for storing in the club library in the village since the club could not afford to buy the books. The club organised football tournaments for which a shield was donated by our grandfather. Shri Meghen (who later became Chairman of the UNLF, a Manipur insurgent group) was one of the players in the club team. He used to stay in uncle's house during these tournaments. I recall the day when my elder cousin Ta Guno received his first pay of a princely sum of Rs 75 as a school teacher. He took me to the market at Thoubal for a 'bora treat'. He asked me to eat as many *boras* (*pakoras*) as possible and even carry some home! I have very fond memories of all the elders in the family as well as in the school who looked after me as their own little brother.

A Journey to the Unknown: Sainik School in Assam

I cleared my entrance examination to Sainik School. The interview for selection for entry to the school was conducted in the office of the Principal, DM College in Imphal. I remember a question put to me by the Principal of the college who was one of the members of the Interview Board. He asked me if I could identify a wall mounted picture of a gentleman in the office. I looked at the picture and had no idea. I counter questioned him on whether he knew the person! There was loud laughter among the members. The picture was that of Mahatma Gandhi, my childhood iconic name! Nonetheless, I was selected for admission to Sainik School Goalpara, Assam. There was no Sainik School in Manipur at that time. That year I commenced my journey in pursuit of my childhood dream, a dream to join the army and become a Brigadier!

I joined the newly-established Sainik School in Goalpara in 1968. My mother was not very much in favour of my joining the army, but my father and other relatives prevailed over her saying that students from the school need not join the army after passing out from the school; and that I could try for any other profession if I chose to after school. My mother was also told by some that military officers do join high ranking posts in civil after retirement. Mr Baleswor Prasad, IAS, an ex-armyman was the Chief Commissioner of Manipur during those days. Manipur was a Union Territory under a Chief Commissioner then. My mother pawned some of her gold to pay for my school expenses. I was the first from the village and the neighbouring area to study outside the state. A large number of cousins and friends came to wish me luck the day I was to leave for the new school. A small feast of fish and chicken was cooked by my cousins.

The first lap of the journey began the next day by a passenger bus to Naorem Leikai near Imphal where I was required to stay overnight in my uncle's house. The bus to Dimapur, the railhead, was scheduled early morning the next day. My cousin Birkumar on a cycle carried my box and another bicycle carried me and my uncle the late Shri Khomei to the bus stop five km away. There were two state passenger buses daily from Imphal to Dimapur railhead those days. The bus commenced its journey at 6 a.m. and drove through the winding and narrow roads of the hills. I came across a place named, Kanglatongbi where fierce battles were fought during World War II. The World War had been fought in the hills and valleys of Manipur and present day Nagaland just 23 years ago. We reached Dimapur late in the afternoon. We halted for the night in a small lodge near the railway station. I heard the frequent whistling of the coal-powered train engines while resting in the lodge. I was very excited at the prospect of seeing a train and travelling in it for the first time in my life. In fact, there were very few in nearby villages who might have seen a train. The next lap of the journey commenced when we boarded the train to Gauhati. The train wound its way through the jungles of Assam. The thick forest and dense, dark woods were unending. I also enjoyed the *muree* (a concoction of puffed rice and chopped onion tossed in mustard oil) sold by the unauthorised vendors inside the running train during the journey. The vastness of the Assam forests amazed me no end. We passed through a railway station named Hojai ahead of Gauhati, where many Manipuris who migrated from Manipur during the Burmese war 1819 are settled.

We were received warmly by the staff of the school at Goalpara. My uncle left me in the school and he returned home. I was totally lost in the new environment. The English language I learnt in the village school hardly helped me to communicate with others. I came across some Manipuri students within a few hours of my arrival and it became easy to settle down thereafter. I found out later that most of the Manipuri students in my class were elder to me. The school was located in the middle of forested area. There were students from almost all the North East states. Arunachal Pradesh, Meghalaya and Mizoram were still part of Assam then. Teachers were knowledgeable, well behaved and certainly spoke better English than my village school teachers.

Settling down in a residential school requires a lot of adjustments, learning and can sometimes be stressful too. Common bathrooms, strict routine, dormitory type living, friends from varied cultural, ethnic, food and linguistic backgrounds made it more challenging and fun too. Strict discipline was enforced by the teachers and the school student representatives. However, there were instances of quarrelling, and even fighting between various groups of students, such as Nagas, Manipuris, Khasis and Assamese. The diversity of tribal friends I interacted with from various parts of Assam, Nagaland and other North East states in the school provided me with many opportunities to learn about the region. I made many friends from other states too. The routine began with morning physical training under strict supervision of Mr Thapa, a retired army boxer, or with drill under our National Cadet Corps detachment instructor. This was followed by academic classes till lunch time. An evening games period of playing games and sports followed thereafter. The study period before dinner was in our respective classrooms supervised by seniors and the duty teacher. We were completely isolated from the outside world. And this routine got repeated till I passed out from the school in 1974 to join the prestigious National Defence Academy at Khadakwasla.

Lessons from the Boarding School

What was my experience in the school? What is the best thing that I imbibed during my years there? What is that I learnt that would stand me in good stead in my later years? These are some of the questions that I faced in various forums in my later years. School days are special. Growing up years in a boarding school environment are difficult to describe. Even more difficult is to ascribe any one quality I developed there. Yes, the school taught us to be independent, groomed us for physically challenging situations, showed us the path to leadership and team work, not to forget the standard of the academic excellence that the school provided us. Self-improvement activity was one special area which I pursued over the years in the school. The seeds of aspiration to excel were sowed during the school years. I was physically weak even though I was in the school hockey and football teams in my last years at the school. I drove myself hard to improve my physical strength at every opportunity. I even did boxing to get rid of the fear of senior bullies in the school. I practised every possible game on Sundays and holidays to learn as many games as possible. It was a dream come true when I found a berth in the school football and hockey teams.

I became an addict of the news time on All India Radio's 8 a.m. bulletin. Besides the information and the news, I learnt to use better words by listening to the AIR news readers. I visited the library regularly to read more and more books. Some of the books I read during the period were *The Guns of Navarone* and *The Longest Day* (both war stories of World War II), *The Bermuda Triangle* and a book on Sardar Vallabhbhai Patel. I particularly liked books on nature and autobiographies of famous people of that time. Many of the books went well beyond my comprehension, but I did enjoy reading them anyway. I knew that I would benefit in no small measure from the lessons of life of great men and women mentioned in the books. I volunteered for several extra-curricular activities to gather more knowledge about things around me. There were simply too many things to explore if one chose to. Within the limited facilities available in the school, I think I gained immensely from the school curricula and from values that our teachers inspired in me.

In the aftermath of the 1971 war with Pakistan, our school authorities organised trips to nearby places where captured Pakistani army equipment was displayed. A Pakistani navy barge which was captured in the Brahmaputra River was also shown to us near the border town of Dhubri. The barge had sailed upstream from then East Pakistan towards the Assam border along the Brahmaputra River for a reconnaissance mission during the war. We visited the Air Force Base at Gauhati, where we got to see the Air Force planes. The motivational speech by our headmaster Major Vyas (now 84 years old in Delhi) was particularly impressive. I enjoyed the various hobby classes too.

There was much infighting and groupism among the Manipuri students. Each group tried to dominate the other. Some of the older boys even tried to bully me into gambling and such like activities. Some of these boys often broke school rules by indulging in drinking. They felt a sense of bravado by faring badly in examinations. This led to expulsion of at least 12 students from Manipur from my class itself on academic and discipline grounds. These students were intelligent but somewhere they lost their way. Perhaps it was juvenile adjustment issues.

There were a few teachers and senior students whom I used to look up to as role models. I admired and envied the manner our Principal Lt Col S.S. Menon and Mr R.P. Singh spoke English. I admired the gentleness of Mr Sonowal, our mathematics teacher. I was looking out for good qualities in others so that I could try to emulate some of their qualities. One quality I envied was the ability to sing but it never materialised. Many of my classmates spoke reasonably good English but I was very poor in speaking English though I used to score more marks in the subject in the written examinations.

I think I did have a streak of rebellious attitude particularly during the formative years in the school. I recollect a very naughty action of mine in the Eighth Standard when there was bait for some sweets from the canteen in return for throwing a Colgate toothpaste packaging box at a particular teacher during his rounds in the class. The teacher was very strict and insisted on punishment to the students for incomplete work. The class made a plan to scare him! I was the class monitor too. I took the bait, climbed to the ventilator above the window and threw the toothpaste box. The missile missed the teacher. I owned up my action of throwing the paper missile to the teacher. He asked me why I did it. I told him that it was bait for sweets and I took the bait. The teacher was most benevolent in pardoning my action and told me to improve myself. The entire episode was a big lesson in life for me. The power of pardon, the impact of telling the truth and the courage to bear the consequences were lifelong lessons for me.

I participated in and enjoyed the Sainik School NCC (National Cadet Corps) camp in the Almora Hills in 1970. The smell of the pine leaves, the joy of walking on the fallen pine leaves in the Kumaon Hills, and the panoramic view of the Himalayas from Kausani in the same district were truly amazing. We had to make do with two mugs of water in a day for washing and drinking. It was during the Eastern Zone Sainik School NCC camp at Puri in 1972 where as a Camp Cadet Adjutant that I felt a sense of achievement. The camp was located near the seashore. I felt a sense of recognition and of greater self-confidence on being made the Cadet Adjutant. Lt (later Lt Gen.) K.P.D. Samanta, a Gunner officer, was the Camp Commandant. Thirty-four years later he became my boss at the Army Headquarters when I joined as Director of Personnel Services in 2005 and he was a Major General. It is another matter that I was relieved from the responsibility as the Camp Adjutant on the grounds that I was unable to exercise control over the 600-odd NCC cadets to prevent them from going near the seaside. It was in the academic field that I maintained the top position in the class during most of my stay in the school till the Central Board examination in 1974. I was the class monitor from time to time. I was made the school vice-captain in 1973. This gave me ample opportunities and the confidence of speaking in public. I learnt to dress better. I started to learn the art of leading and to captain the large number of students in the school.

Journey in a Hijacked Train

I cleared the UPSC (Union Public Service Commission) written examination for entry to the National Defence Academy in December 1973. The interview, psychological and personality tests for the final selection were held at Number 34 Services Selection Board (SSB) located at Roorkee in May 1974. My friend Sorjit and I from the same school cleared the Selection Board. We were returning to the school by train via Lucknow. By the time we reached Lucknow, an indefinite All India Railway Strike led by George Fernandes, later the Defence Minister of India, had begun and no trains were running. Sorjit and I spent two days on the railway platform there. On the third day, we got to know that some people including a few army men had commandeered a few coaches of a train and were to proceed towards Siliguri in West Bengal. At this point in time we had only a one-rupee coin to spend between the two of us. I told Sorjit that we must eat some bread before we commenced the journey. He told me that it was better to keep the money for worse was to come and he took and kept the one-rupee coin with him. We somehow managed to get some space inside the train and decided to face the consequences. The train presumably driven by army men stopped at every bifurcation of the track to ascertain the direction to follow. The train did not run at night whenever there was confusion regarding the correct route. During the day the train used to stop near mango orchards in the vicinity of the rail tracks. We enjoyed the frequent mango breaks. The railway platforms en route were all empty. The train did not stop at any of the stations but occasionally stopped at places where no law enforcement agencies were present.

The train reached Siliguri after six days of very exciting and adventurous journey of a lifetime. The train journey in normal times would have taken 30 hours or so. The army men inside the train treated us well. We impressed them with our SSB documents to prove that we were soon to be army men too. On reaching the Siliguri station, we saw the Military police round up all the military personal who had come in the hijacked train. We were left free to go. Go where and how, was the challenge now! We were 300 km from the school. Luckily, we located a goods truck driver near the railway station who told us that he was proceeding to Bongaigaon on the northern bank of River Brahmaputra. Since our School was on the southern bank at Goalpara, we decided to hitch a ride in the truck. On boarding the truck Sorjit took out the same one rupee coin which he had taken from me at the Lucknow railway station and gave it back to me! We paid the truck driver by selling my wrist watch for Rs 135. We reached the school safely, 11 days after we left Roorkee! The experience of travelling six days in a hijacked train is unthinkable today. Sorjit and I used to reminisce the event very often till his unfortunate demise due to cancer in 2018. It was India of the 1970s. Soon I was called for training at the NDA in May 1974.

Discovery of the Self – National Defence Academy Years

In the British years in India, it is told that Kipling once said, 'If you want your son to be a man, send him to Sanawar School.' In our times, I think there couldn't have been a better training facility than the Academy to make one a true leader of men. The NDA is also aptly called the 'Cradle for Leadership' . The NDA logo is 'Service Before Self'. I joined the NDA in July 1974 along with nine of my classmates from Goalpara. My 11th Standard marks were decent enough to have got me into St Stephen's College in Delhi. My mother once again suggested that I should go to Delhi for further studies. I was, however, able to reassure my mother that I would do better in the NDA and the army. She wasn't convinced but she let me go to the NDA. Though my family was reasonably better off than most in the village, we did not have regular income to provide the best to all the brothers and sister in those years.

Khadakwasla is one of the most picturesque corners of Pune region. This location was chosen by Shri Jawaharlal Nehru, the then PM. The approach to the Academy from Pashan area was most impressive. Well-groomed hedges, clean roads, majestic buildings were a treat to see. Life as a first term cadet in the NDA is difficult to describe. I reported to the Echo Squadron on joining. The first thing I noticed in cabin number 27 was the neatly displayed poetry *If* written by Rudyard Kipling. This poetry perhaps was my best teacher at the NDA! The poetry kept me going in difficult times. Air Chief Marshal B.S. Dhanoa was one of the cadets in the same squadron in 52nd Course. General Bipin Rawat was another cadet in the neighbouring squadron. Being from a Sainik School background made life a bit easier on two counts. One, some known faces were there already. Two, the activities were similar but more demanding and rigorous in nature. I shall never forget the first morning tea. I was detailed to carry tea for a Fifth term Sergeant. I went to the tea room without a dressing gown! It was my first offence in the Academy. The seniors seemed to know that I was a First termer, so they didn't bother me till a Second termer caught me by the ear and ordered me back to the cabin. In walked the duty orderly with a steel canister full of boiling muddy water masquerading as tea. The chaos in the tea room had already begun. The Third, Fourth and Fifth termers were entitled to have a decent cup of tea whereas the Second termers carried the tea for the Sixth termers before they opened their kingly eyes in the morning.

The First termers, I learnt later, were not even to be seen near the tea room! I was the privileged First termer who had been selected to carry tea for the Fifth term Sergeant. Luckily I could get some tea for the sergeant and everything seemed normal till I placed the tea in a wrong cabin not belonging to the same cadet sergeant. All hell broke loose by then. The sergeant sent for me and I landed up rolling, rolling and continuously rolling on the ground for what he described as a big offence of 'doping' not pardonable by NDA standards.

The routine was almost on the lines of our days in the Sainik School. More physical activities, far superior infrastructure and many more opportunities to learn. However, we found it impossible to avail of the vast opportunities available at the NDA in terms of sports, hobbies, library and other curricular activities due to lack of time. The focus of our routine remained informal physical activities. There existed a very strong Squadron spirit to win the various sports and professional competitions. The squadron was the ultimate in our closed environment of the NDA. Formal training in academics, physical training such as PT, horse riding, swimming, drill, hobbies, running and other routine activities were enjoyable. But the informal squadron activities such as cabin cupboard maintenance, extra running practices at night after dinner, sometimes inside the building itself in order to win the inter-squadron cross country competitions, extra physical training during Sundays and holidays as also after-dinner physical activities were all directed towards winning the championships. Coursemate spirit, squadron spirit were the most important issues around which our lives in the NDA revolved 24X7. Getting things done by hook or by crook, being smart enough not to get caught while pursuing these survival techniques, slowly became a part of our lives. Interest in academics was the pursuit of a few.

It became clear to me as days went by that individual growth took a backseat. Among the 1,500 odd cadets, the individual winners in various competitions were far and few. Very few systemic programmes were available to enhance individual skills and personality development. Therefore, I needed to reorient my efforts while undergoing training in the academy. The aim was to enhance my interests and my strong points. I dedicated more time in the library whenever possible. I joined hobby classes such as astronomy, nature lovers club. Many cadets mistook me to be a foreign cadet. Most North East cadets were branded as Nagas by the other cadets, just as cadets from the south were called *Thambi*. I often wondered why this branding. Perhaps, due to our looks. Maybe due to our own behaviour and demeanour. It certainly was not a derogatory or racist comment. NDA taught us how to laugh at oneself. I discovered that all my confidence in public speaking gathered from the school did not help in the NDA. I continued to practice speaking with correct pronunciation off and on to improve my communication skills. I represented the squadron in hockey, football and boxing. However, I was nowhere in the consideration for the Academy team levels. Development of Officer Like Qualities (OLQ) was an important component of our training. The same was achieved through counselling and interaction with the squadron officers. Cadets usually look at the conduct of the many officers who trained us to pick up threads towards our own personality development in some way.

Life in the academy wasn't without some crazy moments. On one occasion, I took the step of throwing the bicycle belonging to one of the seniors into a nearby well. This was to avenge the ill-treatment he meted out to me frequently over minor issues. But the risk involved in the act, if caught by the authorities, was considerable. On another occasion, the seniors sent me and another friend Cadet Routela to the lawns of majestic Sudan Block to bring a few ornamental fish and pluck a few lily plants from one of the garden ponds to adorn the squadron aquarium without getting caught! Luckily, we were successful on our mission! There were such activities all around. Some course mates raided the canteen to lift sweets from there. Many cases of minor disobedience of orders happened all the time in every nook and corner. These were considered naughty activities, otherwise punishable under the Academy regulations. Most seniors looked the other way except for the Cadet Appointments and the Drill 'Ustad' (instructors) who were mandated to enforce discipline in the academy under the vigil of the Academy Adjutant. It was near impossible to check all the wrongdoings all the time everywhere. And many cadets took to the cheap thrill of disobeying orders and feel great about it. I reckon, such activities were mostly to get over the setting in of boredom and a false sense of bravado.

Life in the otherwise cocooned environment had many other attractions too. The Gole Market where eatables were available was a great place to visit. My pocket money was a sum of Rs 15 a month. The amount could provide me some extra eats from the market. This sum had to meet the payments to the Bearer who polished my shoes, made my cabin neat and clean and attended to other chores. The remaining money was spent in purchase of toiletries, personal needs, Rs 5 contribution to the squadron funds and some others. Two movies in a week gave us adequate entertainment. The Academy auditorium was a comfortable place for us to sleep with no one to bother us. Film stars from the Film and Television Institute of India in Pune occasionally came over to perform in our auditorium. Nothing could beat the weekly cold coffee and pastries on Saturdays during evening tea time. Tipsy Pudding during camp fires of camp Greenhorn, Camp Rover and Camp Torna are unforgettable times of any ex-NDA. Any get together of ex-NDAs will be incomplete without mention of Tipsy Pudding. It was normally made on the days of Guest Nights, and special occasions like the outdoor camps. During the various camps it was customary for cadets to hijack maximum Tipsy Puddings, pack them and send them to the squadron seniors in mess tins. In fact, the encounters between Cadet appointments and ordinary cadets inside the campus was a veritable sight of entertainment. Not to forget the sight of 7-10 metres jumps in the swimming pool where the willpower of the reluctant jumper cadets was tested! Some call the 7m jump one of the most exciting thrillers on the earth. One could predict the behaviour of each reluctant cadet on the 7m board. One hesitant step forward, one step backward, and it would go on till he was pushed by the Ustad or a senior into the swimming pool.

The thrill of running the cross-country race at the Gliderdrome where 1,500 cadets were flagged off at the same time made for unforgettable moments. Horse riding where the horses usually overpowered the greenhorn riders was a sight most entertaining in the equitation area. For some cadets it was seldom that they could be on the horseback itself. The breakfast break on a working day, particularly near the Ashoka Pillar, was perhaps one of the most exciting scenes ever where nearly 1,500 cadets dispersed in a matter of a few minutes in cycle squads from the training areas to their rooms, to change, have breakfast in the cadets mess, pick up their bicycles, proceed to their next training session in a matter of a little over 30 minutes. All this under the watchful eyes of the drill Ustads and the Cadet Appointments who were watching with hawk eyes to catch the defaulters of the academy regulations and discipline. Some of them used to check the turnout of the cadets, some checked haircuts. Riding a cycle 'out of squad' was a serious enough default to get a 'Sinhagad' hike during the weekend.

Sinhagad is a Maratha hill fortress which is famous for the actions of Shivaji against the Mughals. The hillock overlooks the Academy from the northern end of the Khadakwasla Lake. One of the minor punishments was a trek to the top of the Sinhagad fort and back within a time limit of about six hours with a full field service Marching Order Pack 08 fully loaded (a fully kitted outfit of a soldier). There were formal and informal punishments. Formal ones were meted out by those authorised by the Academy rules, and informal ones were those unauthorised punishments meted out mostly by unauthorised seniors or unauthorised punishments by authorised appointments. Unofficial ones like 'Seventh Heaven' (jumping as high as possible and holding on to the mesh wire), 'Bajri Order' (army kit filled with gravel), 'Rolling' (rolling on the ground), 'Lifting Cycles' were frequently meted out to the cadets for minor misdemeanour type transgressions. Such misdemeanours could be bad drill, improper turnout, out of squad movements or even giving dirty looks to the seniors!

The team spirit, camaraderie and life skills acquired during the time in the Academy has stood most of us in good stead in later years of the profession of arms. The will to perform the task under adverse conditions and not to give up under any circumstances were perhaps the best lessons we imbibed.

In retrospect, I feel that the training in the Academy could have been more fruitful had the syllabus and the curricula been modified in keeping with the need of preparing future leaders. An attempt of such nature was made in 1972 when the entry level to the NDA was raised to 11th Standard and the cadets passed out from the Academy as graduates in Arts or Science stream. It brought a quantum change in terms of academic progression, but I am of the opinion that the much-needed grooming of the future leaders remained unaddressed to some extent. A cadet needs to be 'selfless' first and foremost. The need is to transform thc DNA, to become 'selfless'. Age-old traditions of ragging, occasional cases of informal punishment of extreme nature amounting to sadistic behaviour among some cadets got the better of the institutional needs. Mental and intellectual growth remained static. In 2001, I was posted as a Battalion Commander No. 4 Battalion of the NDA after completion of my Infantry Battalion command. As a Battalion Commander of the cadets, I tried to bring up the issue of inculcating leadership qualities, understanding how to lead men, introduction to management issues and ethical issues to some extent. However, due to my short tenure there and very compact JNU(Jawaharlal Nehru University) academic syllabi, not much could be achieved. More about this tenure at the NDA later in the book.

My three years of pre-commission training in the Academy were relatively smooth. I achieved modest academic standards, took part in most sports activities, and picked up some good hobbies and lifelong interest in reading and writing. I was appointed Divisional Cadet Captain in my Sixth term. The passing out parade on 7 June 1977 was the pinnacle of our young lives when we graduated from the Academy to join the respective service academies (Indian Military Academy for the army) and to be commissioned in the forces after an additional year of training. A spectacular drill by nearly 1,200 cadets followed by an IAF flypast to bid us farewell was a memory of lifetime. My uncle the late Shri Khomei and my younger brother Nongyai attended the ceremony. With a heavy heart, I along with all 244 of my coursemates from the Army, Navy and Air Force bade farewell to the alma mater that day. Looking back to see the dome of the Sudan Block as we headed towards the Pune Railway Station in the Academy bus was a memory of a lifetime. Nearly 40 years later, on 16 December 2016, many of us from the same course along with our spouses got together at the NDA campus to relive our glorious past. Some of us in the three -star ranks, including Gen. Bipin Rawat, the present Chief of Defence Staff, and Air Chief Marshal B.S. Dhanoa, were among the serving during the memorable event.

The Antim Pag (The Last Step to One Star)

The Army Cadets proceeded to the Indian Military Academy, Dehradun in July 1977 for the final lap of army training before getting commissioned. The routine at the IMA was similar to that of the training at the NDA. However, the charm of team spirit and bonhomie among the trainees was missing. Most cadets took to doing better to be in higher merits. The training mainly comprised of professional subjects to enable us to command an infantry platoon. Here, I came to meet many officers of great repute. Gen. Shankar Roy Chowdhury was our Battalion Commander in Manekshaw Battalion then. He had a remarkable memory. Once I asked him how he could remember everything that he saw or heard. He politely told me to pay attention to what I saw, heard or read to retain the things in memory. I could sense during the one-year training that the competition among the Gentleman Cadets was becoming more and more intense. There was greater will to perform and do well in every subject of training. I took things as they came. Worked hard, not to outdo others but to improve myself. My Company Commander once told me to be more assertive. I told him I would try. But I knew that I would continue to be the same. After one year of gruelling training, we graduated as Second Lieutenants on 10 June 1978, the day we became officers of the Indian Army. I was allotted Infantry, the Second Battalion of the Rajput Regiment. I was 21 years and a few months when I got the commission. We took oath in the famous Chetwode Hall after stepping out for the Antim Pag (The Last Step) at the entry from the parade ground into Chetwode Hall. I knew that I would aspire to adhere to the IMA Honour Code written by Field Marshal Sir Philip Chetwode.

> *The safety, honour and welfare of your country come first, always and every time.*
> *The honour, welfare, and comfort of the men you command next.*
> *Your own ease, comfort and safety come last always and every time.*

It is with this priceless wisdom and my resolve to adhere to the same that I bid adieu to the Cadet/Gentlemen Cadet (GC) days and to face the challenges of the real world as an officer in an infantry unit.

East Meets Soldiers of the West

The day we were eagerly waiting for so many years came true on that day. I was IC 35206W 2/Lt KH Singh of 2 RAJPUT. The unit had earned the distinction of fighting all the wars in India and abroad since its raising in 1798. We were four from Manipur to be commissioned that day. Khulem got 2 SIKH, Debendra was allotted the Ordnance Corps, Bir Mangal got Army Supply Corps. My school and classmates who joined the NDA from Goalpara on the same day also got commissioned. All of us headed home together in the train. Back at Charangpat, my grandparents, my parents, brothers and sister, my relatives were happy to see me as an officer. My grandmother was quite amused by the various types of service uniforms I got in my baggage.

I joined 2 Rajput at Poonch in July 1978. The unit comprised of troops from the Rajput, Gujar and Bengali community. The unit Commanding Officer was Lt Col S.N. Verma, with whom I maintain contact to this day. I was assigned to the A Company under Capt. V.K. Singh (who later became General and Chief of Army Staff). The Company was deployed along the Line of Control (LoC). My tenure under Capt. V.K. had significant impact in my growing up years in the unit and beyond. At that time, Capt. V.K. had just returned from a Rangers course in USA. He provided me many books and military précis to study. During one of our visits to another satellite post along the LoC, we lost our way and landed right in front of a Pakistani heavy machine gun bunker. We perhaps had strayed into the other side by a few metres. Luckily, the Pakistani NCO (Non-Commissioned Officer) manning the bunker did not open fire, instead barked at us with the choicest of abuses in Punjabi for us to fall back or else.... Our party was totally caught by surprise and the right thing to do was to fall back with grace. This we did.

Capt. V.K. was a physically tough officer and had been a topper in his batch. He also guided me to be the best among my troops in every sphere of professional activities. Be it running, climbing, weapon handling, minor tactics or any administrative assignment. The 2 Rajput was also named *Kali Chindi* during their actions in the Burma Campaign of 1944. In Independent India, the unit took active part in the liberation of Poonch and Rajouri during 1948-49 under the able command of Col (later Brig.) Ranbir Singh. The most important action in the history of the unit was of course the Battle of Namka Chu during the Indo-Chinese conflict in 1962. On 20 October 1962, an entire Chinese division with over 12,000 soldiers swiftly advanced through the solitary 7 Brigade companies and platoons strung along the Namka Chu River or nearby in a mostly non-tactical manner. In a matter of a few hours 7 Brigade was overrun after heavy fighting. Lt Col M.S. Rikh, Commanding Officer 2 Rajput, along with 90 others, was taken prisoner while 282 brave soldiers fought to the end. Only 62 men of all ranks belonging to the unit mostly from the administrative details got back home. Major B.K. Pant's last words were 'Men of the Rajput Regiment, you were born to die for the country. God has selected this river to stand up and fight the Chinese and die.' Of his Company's 112 men, 82 were killed. His son Mukul Pant joined the Army Medical Corps and became a dear friend. He eventually retired as a Brigadier though he deserved much more. I had the privilege of serving with many NCOs and JCOs (Junior Commissioned Officers) who fought in 1962 in my A Coy (Alpha Company). Hav. (Havildar) Saudagar Singh was one such soldier who captured an LMG (Light Machine Gun) from a Chinese soldier and killed as many as 15-20 Chinese soldiers. We heard many stories from the survivors. I learnt valuable lessons from them.

Life in the LoC was exciting and challenging. The Simla Agreement between Smt Indira Gandhi and Mr Zulfikar Ali Bhutto in 1972 after the defeat of the Pakistani army in East Pakistan resulted in the LoC to demarcate the temporary boundary between the two countries. The actual demarcation of the LoC was not easy to transpose on the ground at many places as the survey instruments were not very accurate then. Both sides had their perception of the LoC on ground. Frequent firing had to be resorted to guard and prevent transgressions which both the sides believed to be in their own territory. Many minefields had been laid during the earlier wars of 1948, 1965 and 1971. These minefields, particularly the ones laid during the 1971 war which had taken place just seven years ago, made our movements extremely restricted in the area. Frequent drifting of mines on the mountain slopes made our movements trickier.

The Simla Agreement between the two countries forbade either side from constructing new defence works or improving the existing ones. Any violation by either side was immediately reported to the authorities and suitable actions were taken by them. Rations, defence stores, personal kit, water supply, ammo and all maintenance items had to be ferried on the back of troops. Very few porter facilities were available during that time. The challenges of operating in the LoC largely remain the same in the last 40 years of my service during which I served in the LoC for close to 10 years at various periods of time. Post 1990s, the challenges have increased manifold due to the threat of infiltration of Pakistani trained militants from across the LoC.

As a Lieutenant, I was required to go on patrolling and ambush duties in the vicinity of the LoC quite regularly. But my company commanders always ensured that senior NCOs accompanied my team most of the time. As a Second Lieutenant, with one star on my shoulders, I thought I was ready to take on the world. My seniors had a different view though. To them, a Second Lieutenant has many more things to learn to be able to command a small body of troops. I was made to learn about each and every detail of the soldiers of my Platoon and the Company, including details of their family members. The four-month-long Young Officers (YO) course, including the Commando course at Belgaum, taught me that the human body can withstand any physical hardship to any limit under most trying circumstances.

Untimely Loss of my Sanawarian Brother

During the YO (Young Officers) course in Belgaum, I learnt of the sad demise of my younger brother Khogen. He was a gem to me. He was the most gentle and intelligent among all of us. He was in 11th Standard in Lawrence School, Sanawar, Shimla. He was suffering from Systemic lupus erythematosus (SLE), an autoimmune disease, while in the school. I had gone to Sanawar during October 1978 to take him to a better hospital in Delhi for treatment. Later I dropped him at home in Charangpat. My brother did not know that it was going to be his last days in Sanawar. The school gave us a van to have us dropped at the railway station. He preferred to trek down 2 km to a place called Dharampur on the highway to Kalka. We were accompanied by his classmate Sukhbir. On the way, Khogen asked me if it was possible to own a home in the hills near Sanawar. I told him that I would think about this at an appropriate time. That dream of his never materialised though. I never divulged to my parents that doctors in Delhi and Chandigarh had told me the fact that Khogen was not going to live long. It came as a total shock to me but I decided not to let my parents know this sad fact. And I proceeded to attend the Commando course at Belgaum. It was one of the saddest events in my life. He was very close to his classmate Sukhbir Badal, ex-Deputy Chief Minister of Punjab. Sukhbir was always there by his side when I went to Sanawar. Khogen once gifted me an expensive trouser. I asked him where he got it from. He replied that he had got it from Sukhbir for a sum of Rs 2. He also remarked that the rich have no value for money. I had put the education of my brothers and sister as top priority. My parents, who never went beyond the Fifth Standard, appreciated the importance of education and they did all they could for our education in spite of our meagre income from the paddy fields. My other siblings—Nongyai, Santosh, Kiran and Ichan—continued their respective educational pursuits in the village at that time. Life goes on.

A Life of a Lieutenant!

There is a saying in the army, 'Never believe a Lieutenant with a map!'. I was determined to disprove the saying. As a Lieutenant, my foremost duty was 'To know the Men', and 'Know your Job'. Seemed simple in the beginning. It involved an unending process of learning and knowing everything about each soldier in the platoon including his habits, origin, their belief system and many more. To know the job meant achieving mastery of every weapon and equipment and tactics involved in every facet of warfare, not forgetting to be able to do a hundred other physical activities better than anyone in the group. One had to be quite familiar with every regimental institute, such as unit *mandir*, canteen, Quarter Guard, JCOs mess, etc. The most important was to learn and be part of the traditions, culture, origin, history of the men in the unit. We had 50 per cent Rajputs from North India and 25 per cent each of Gujjars and Bengalis. Learning the dialect and folk songs of the Rajputs and Gujjars were top priority for me. Learning the Bengali language was essential to command the Bengali Company. It was mandatory to memorise the names of all the Commanding Officers and the Subedar Majors of my unit since its raising in 1798! This was to ensure that the regimental linkages of the unit personnel continued. The troops accepted me earlier than I thought I would be. Regimental traditions demanded that troops must accept the officer if one has to command them in war. My company comprised of Rajputs by caste from Rajasthan, Haryana, Uttar Pradesh and Madhya Pradesh.

Even though they were from the Rajput community, there were visible differences in their looks, habits and social customs. As a young officer I had to be very mindful and sensitive to these issues while dealing with my soldiers.

During one of the routine maintenance days after a heavy shower of rains at an isolated post, I gave orders to the company senior NCO to have the track inside the post properly maintained. This happened three months after I joined the company. I did not see much of activity to carry out the said order for some time. I waited for a few more hours and inquired with the senior NCO if my order was clear to the troops who were assigned the task. He hesitantly replied in the positive. Yet nothing moved for another hour. I decided to gather the troops and have the task completed as per my orders. The NCO promptly came to me and wanted to convey something to me. He inquired with me when was I due to leave the unit, thinking that I was from the Royal Bhutan Army (RBA) whose officers came for attachment to our unit for three months in the earlier location. Now I understood why his responses to my orders were quite muted. He was under the impression that I was from the RBA and RBA officers never gave any orders to our troops! Perhaps the troops made the wrong assumption due to my mongoloid looks. Slowly but steadily I worked myself to win the trust and confidence of the troops under my command.

On my return from the Commando course, a unit level training exercise was held during December 1978 in the snow-capped mountains of the Pir Panjal Range. I was to command a platoon size force in a defensive and offensive operation. It was during an attack exercise that I learnt about the grit and determination of our jawans. Under heavy snowfall, the unit personnel climbed the 14,000 ft high feature in record time at night with full battle load. Later, during my tenure as the General Officer Commanding (GOC) and as a Corps Commander in the same sector I asked my pilot to fly over some of these mountain peaks under whose lap I had dug trenches, weathered the snowfall and the blizzards and above all slept under the clear skies in the icy nights as a Lieutenant. Of 16 officers who were in the unit at the time, nine went on to be Brigadiers and of them four became Generals in the later days. A remarkable distinction for any unit.

On one occasion during my young days in the unit, the Commanding Officer (CO) asked for volunteers for a long-range patrol across the Pir Panjal Range from Poonch to Gulmarg. I volunteered and so did my Company Commander Capt. V.K. Singh and a few other officers of the unit. The CO decided that I was to be detailed since I was the junior-most. This patrol was one of the most memorable experiences of that time. The party of 10 officers of the formation along with their buddies set off from the base of Pir Panjal to cross the mountain ranges in Chor Panjal Pass and later trek to Gulmarg in about 3-4 days. The beauty of the Pir Panjal Range was most captivating. This enthralling experience of high-altitude trekking made me volunteer for more mountaineering and trekking trips in the Himalayas later in my life.

I made every effort to read as many books as possible. I also tried to study the flora and fauna of the foothills of the Pir Panjal Range. A book titled *The Book of Indian Birds* authored by Dr Salim Ali was extremely useful in my study of birds. Dr Ali had come to NDA once during my hobby classes of 'bird watching' there. My stay in the higher reaches of the mountains gave me ample opportunities to study the mystery of stars—my another hobby. The 'Astronomical Club' at the NDA gave me some idea about the mystery of the sky. The hobby classes that I attended while at NDA came very useful in these circumstances. I developed the habit of sustained reading due to the encouragement of my seniors like Brig. A. Kapila, Gen. V.K. Singh, Lt Gen. Thomas Mathew and my first CO Lt Col S.N. Verma. My interest in reading books on nature increased manifold.

As a young officer, I was also required to learn about the unit routine and master the activities of troops at the section, platoon and company levels. For this I had to be on the ground staying with troops in their bunkers, carry out sentry duties at night, perform duties as a Platoon NCO, Kote (where weapons are stored) NCO, mess commander and so on, besides to refine the weapon handling of all available types in the company. The biggest calibre weapon then was the 57 mm Recoilless gun! This weapon is of World War II vintage, since discarded by the army. In fact, except for the 7.62mm Self-Loading Rifle (SLR), all other weapons and equipment were of 1950 vintage. I recall that we were still using the WWII sound-powered line telephone in many places. One was also expected to master the regimental history in detail. Knowing the history of each of the silver pieces in the Officers Mess was of paramount importance. Through these silver pieces which were displayed immaculately in the unit Officers Mess, we could learn more about the history of the unit. Our Officers Mess was elegantly done. The silver pieces of the 1798 raising of the unit and centenary silver of 1898 were most impressive. The photographs of Maharaja Himmat Sinhji and a painting of Maharana Pratap on his horse Chetak adorned the ante-room.

During November-December 1979, we held an Inter-Battalion Gen. Chaudhary Battle Fitness Competition of all the commando platoons of the entire Division. There were 19 infantry units, including one of the Para Commando units stationed nearby. I was given the task of leading the unit commando platoon to take part in the competition. We trained for a few months prior to taking part in the event. The competition involved a 26-km-run with full battle load, negotiate a battle obstacle course followed by firing of personal weapons in the firing range. I knew that it was a very tough competition. On the day of the run, my CO Lt Col C.M. Chowdhri came to motivate the Commando platoon. We commenced the 26-km-run with the chant of our unit battle cry *Jai Bajrang Bali ki Jai*. Each one of us had to carry 21 kg of battle load and the personal weapons in addition. About 500 m short of the finishing point, one of my jawans fainted and collapsed. It was extremely important for all troops of the platoon to cross the finishing line together as a team, otherwise huge penalties were imposed. I immediately picked up the battle load of the said jawan and told the team members to carry him to the finishing point. Thus I carried two loads of 21 kg each making it 42 kg total. I weighed only 46 kg then. Our team came first in the running part of the competition. However, we could not do equally well in the firing. At the end of the day I had massive skin bruises oozing with blood. This one incident proved to be a blessing for me since the entire unit came to know of my taking on additional load which equalled my own weight and the platoon secured first position in the division in the run.

The officers have to be deserving of the troops they command. Mixed caste units have many good points as well as minus points. The dynamics between the different castes play a significant role in every activity of the unit, including competitions and even war conditions. It is the officer cadre who have to ensure that these dynamics are put to gainful use in operational conditions. Inter–company competitions brought out the best among the troops. They took extreme pride in winning these competitions. However, due to commitments in field and operational areas, competitions were restricted to the minimum. These inter-company competitions tend to throw up caste rivalry at times. This must be discouraged and nipped in the bud at all costs.

During the late 1970s, the LoC was largely peaceful except for sporadic firing to deter civilians crossing the line from both sides. A few villages from either side had their water point on the other side of the LoC. There were family ties on both sides of the line. The Pakistani troops often resorted to triggering jungle fires from their side to detonate the mines and bombs of 1971 vintage near our posts. In sum, each side was trying to improve their own defensive positions in spite of the Simla Agreement of 1972 which prohibited strengthening of defence on the LoC. The Pakistani troops operated mostly in civil dress and mixed easily with the villagers right on the LoC itself. This made our task of identifying the Pakistani troops in the forward zone difficult.

It is said the initial grooming of the young officer determines his future as an officer. Officership in the army was undergoing gradual shift in those days. The tradition of the royalty and highly connected people joining the army was a thing of the past. Young people coming from English medium school background and upper middle class families were the majority in the post-Independence era. Consequently, the 1960-70s also saw the beginning of the Sainik schools and Central schools sending considerable number of officers to the defence forces. This period saw gradual transition in the background of the officer cadre shift to the middle class and lower middle class families. I was fortunate to meet soldiers who had fought in 1962, 1965 and 1971 wars. Their perception of things during the wars taught me many valuable insights of soldiering. We had very professional and spirited officers in the unit.

Great Things Happen When Men and Mountains Meet

It was during this period that a regimental expedition was to be launched to climb Mount Kedarnath and Kedarnath Dome (22,400 ft) in the western Garhwal Himalayas during April 1980. The expedition was being led by the Rajput Regimental Centre Commandant the late Brig. (later Maj. Gen.) P.L. Kukrety. We were 16 members from the regiment. Capt. (later Brig.) Samarjeet of 6 Rajput and Capt. (later Brig.) Jamwal of 9 Rajput were other officer members. This expedition commenced from Gangotri with the trek to Gaumukh, the origin of river Ganges. We were equipped with very simple and rudimentary boots and high altitude clothing and equipment, mostly borrowed from ITBP (Indo-Tibetan Border Police) units nearby. But the will to accomplish the mission was in all of us. Brig. Kukrety was the driving force behind such a team spirit. He was a leader in the true sense. He never rested till the day's work was done. He was widely known in the army for his strong views on contemporary issues in the army. It was really captivating to see the peaks of Shivling en route to our base camp in the Kedarnath Glaciers. I was to lead the climb to Kedarnath Dome, Samarjeet to lead the climb to Bharat Kuntha peak nearby and Jamwal to climb the Kedarnath Peak. My first exposure to real mountaineering in very high altitude thus began with the blessings of Lord Kedarnath from the base camp on 15 May 1980. And it was on 23 May that my party scaled the Dome which stood at 22,400 ft. I was simply elated to have achieved this feat without any harm to my team. There were a few instances of very dangerous ice walls which we had to negotiate while climbing. Luck was on our side that we were able to negotiate without any mishaps. We could hear loud noises of avalanches all around. The sheer magnificence of the mountains made me forget all odds.

On our return to Gangotri and back to Fatehgarh, Samarjeet and I realised that we were suffering from severe chilblains (small eruptions caused by the inflammation of tiny blood vessels after repeated exposure to cold air). We were evacuated to the Command Hospital at Lucknow for further treatment. We remained there for almost a month. There we met Capt. Bahuguna, another keen mountaineer who had earlier climbed Mt Kanchenjunga in 1977. He was admitted there due to frostbite after an expedition to Nanda Devi. The lessons that I learnt from such adventure activities were truly priceless. The spiritual experience in the lap of the mighty Himalayas as we trekked to Tapovan and onwards to Gaumukh (approximately 14,000 ft) where river Ganges actually originates was beyond description. Thus began my tryst with the Himalaya Mountains soon after my commissioning as a Lieutenant in the Indian Army.

On completion of three years in the Poonch–Rajouri Sector, we moved to Lucknow for a peace tenure. Peace tenures are provided to the infantry units to train for war, rest, recoup and look after the family affairs as well. The tenure at Lucknow was also an experience of self-discovery and evolution. Moving into a big city after three years in the mountains of Poonch was a cultural shock in many ways. Besides the usual unit routine, there were many extraneous social activities both in and outside the unit. I made many mistakes during the process of adjustment to the new way of life. As the Adjutant of the unit, I sent a vehicle to drop an air force officer at his base 30 km away at midnight without informing the CO. He pulled me up the next day.

On 30 November 1981, the unit Raising Day was graced by Lt Gen. 'WAG' Pinto, the Army Commander and the Colonel of the Regiment. The event was also to be a farewell for Gen. Pinto who was retiring the following month. Brig. Ranbir Singh (Retd) who commanded our Paltan twice also attended the same function. Both the officers had fought the Japanese forces during World War II in Manipur/Naga Hills as part of 1 Rajput, now 4 Guards. Brig. Ranbir Singh earned his Military Cross in the battle of Kohima. I was assigned the duty of looking after both of them at the function. On his arrival at the railway station in Lucknow he inquired from me if I could arrange a similar tie that may be worn by Gen. Cariappa, former COAS (Chief of the Army Staff) if the General was attending the function. It was a big lesson for me that even after 15 years of retirement he was concerned about his turnout in front of his ex-boss! Since Gen. Cariappa was not attending the issue was resolved, but I had learnt a great lesson. I sensed that both Gen. Pinto and Brig. Ranbir were high on whiskey. But they would not stop. Past midnight Gen. Pinto desired that the elephant waiting for his farewell be brought in. Gen. Pinto mounted the elephant and I was also asked to be with him on the back of the animal. A last drink was ordered for him while being mounted on the animal. He insisted that a drink be served to me as well. After a few minutes, the General slipped down from the elephant. All present at the event were stunned. I jumped from the animal and tried to rescue him. The General was safe and sound. My Brigade Commander was unhappy about the incident. He even suspected that I may have caused the General to fall. I tried to reason with the CO that I was not at all responsible for the incident. The dinner party got over and I was told by the CO to be present in the office of the Brigade Commander the next day at 8 a.m. Reasons were obvious. I was perhaps being admonished for an offence I never committed.

I was present at the appointed time at the Brigade Headquarters. When I entered the office I found that the Brigadier was warm and he did not seem to be unhappy with me. He spoke to me about unit activities and let me go. I was wondering what could have changed his mind overnight. I later learnt that Lt Col Gurbaxani who was in the personal staff of the General had got wind of the intention of the Brigadier to take action against me for what he thought was a misdemeanour on my part to have let the General fall from the elephant. Col Gurbaxani apparently spoke to the Brigadier and told him that the Army Commander desired that no adverse action be taken on this incident and that it was no fault of mine. I felt a sense of relief but I knew that I should be more correct while dealing with such situations.

On the advice of Gen. Kukrety, I enrolled as a life member of the Sky Diving Federation of India (Life Member No. 41). I got an opportunity to go for a skydiving trip to Agra. The skydiving was being organised by the Parachute Brigade stationed there using a Beaver aircraft. It was a momentous opportunity to have carried out six freefalls during the 20 days leave. I had a candle formation of the chute during my last jump. I was dropped from about 9,000 ft. When I opened the parachute using the Rip Cord Handle at about 4,000 ft, I had not noticed the incomplete deployment of the parachute due to which I was plummeting towards the earth much faster than normal. It must have been about 700 ft from the earth that I heard many people shouting 'emergency… emergency' from below. I could open my emergency chute in the nick of time and I was saved from a very bad accident. I have never jumped thereafter again. My freefall skydiving jumps gave me the confidence to face fear of heights. Agra is home to the 'Glass Eaters' as they called the Paratroopers! My friends there insisted that I join the gang of 'Glass Eaters' by chewing glass after my last jump. Not to be outdone, I readily learnt the art of glass eating with blood oozing from mouth. Bravado and self-ego of a young officer!

My one-year stint in Lucknow was cut short by my sudden posting to the Regimental Centre in Fatehgarh (UP). I was a bit taken aback. I was keen to continue in the unit. I requested seniors to help me remain in the unit. I was told that the Brigade Commander had persuaded my CO to have me transferred from the unit. I was transferred to the Rajput Regimental Training Centre at Fatehgarh in January 1983. However, fate had other plans for me. I was placed under Lt Col (later Brig.) Kapila who was my first Second in Command in 2 Rajput. He was also Adjutant of the unit during the 1971 war. I was to perform the duties of Weapon Training Officer and that of Company Commander of the C Coy. As I assumed my duties, I started liking the responsibility to train young recruits to grow into young soldiers of the regiment. Lt Col Kapila treated me with care and almost like a younger brother. Brig. Thakur Ranbir Singh (Retd) visited the Regimental Centre during one of the Biennial Conferences and expressed his desire to gift me a Japanese Samurai sword (*Katana*) which he had recovered from a dead Japanese Brigadier during the Battle of Kohima in 1944.

Brig. Ranbir Singh was the Second in Command of 1 Rajput which took part in the battle. He earned a Military Cross in the same battle. It was a singular honour for me to have been chosen by the Brigadier to be presented with the Samurai sword during a special parade. I still have the sword in my house and it occupies a pride of place in the house at Imphal. The Japanese Brigadier had apparently committed Harakiri with the sword after the defeat in the battle of Kohima.

The location of the Regimental Centre is also a very historic site overlooking the mighty Ganges. The present cantonment fort was established by the British during their march to Delhi along the Gangetic plains in 1777. The first fort in this location was built by Nawab Mohammed Khan in 1720. The fort was repeatedly attacked by the Marathas during the Anglo-Maratha wars. I shall never forget the sunset view across the Ganges as I sat near the bank of the sacred river.

'We have not heard the last of the Manipuris,' Maj. Gen. Sir J. Johnstone, 1893

Meanwhile, my state was going through a series of violence in the valley areas at the time during the early 1980s. The hills in the northern part of the state were already burning since late 1950s as a fallout of the insurgency in the Naga Hills in Assam spearheaded by Mr Phizo. Many Naga tribes of my state (Manipur) soon joined the Naga insurgency movement in the hope that Nagas from various parts of the country and from Burma (Myanmar) will one day unite and be part of 'Nagaland for Christ'. Similarly, the insurgency in Lushai Hills had broken out in the district on 28 February 1966 under the leadership of Mr Laldenga, a former employee of Assam Rifles. Both these insurgencies were actively supported by way of providing training, weapons and base by China and Pakistan. Heavy fighting between the army and these insurgent groups resulted in untold miseries to the common people in the region. The insurgency in both the Naga Hills and Lushai Hills of Assam affected Manipur due to ethnicity factors. The Naga Hills and the Lushai Hills were part of the erstwhile state of Assam and partly excluded from mainstream India due to various exclusive policies of British India such as the Bengal Eastern Frontier Regulation Act of 1873. These areas were mostly left alone during most part of the British Raj since there was little of British commercial interest and nor were these areas considered as strategically important at that time. British strategic interests were served by keeping the kingdom of Manipur as an ally in their Eastern Frontier. Assam kingdom was already in their control. The villages in these mostly tribal areas were allowed to be loosely administered by their respective chiefs as per tribal customs.

In this socio-political setting it was inevitable that the youth of Manipur also took up arms against the Indian State to fight for what they called legitimate rights of the people. Thus, the valley blew up in flames almost at the same time as I was commissioned in the army. The idea of Meitei nationalism was consolidated further due to the emerging Naga polity in early 1950s. The ethnic sub-nationalism of the Nagas also gave impetus to the national identity of the Meiteis, even though Shri Irabot, a communist leaning leader, was the first to fight against monarchy and for an independent Manipur. The Armed Forces Special Powers Act 1958 (AFSPA) was imposed in the entire state in 1980. There were massive cordon and search operations, frisking of people and vehicles in the state. The cadres of the insurgent groups were a very motivated lot. Perceived neglect of the Centre towards the state, historical legacy and unemployment were the major factors for the youth to join the insurgency movement. These were also the days of Kung Fu in Manipur wherein martial arts became a craze among the young people. One police officer of repute told me that Bruce Lee in the movie *Enter The Dragon* was one of the motivating factors among the youth to learn Kung Fu fighting. Western movies of the Wild West did attract young minds towards the gun culture.

I saw thousands of people being subjected to searches and checks under the umbrella of AFSPA. The searches were considered necessary since many innocent young boys and girls were being used as couriers by the militants. On the other hand, the conduct of the soldiers involved in these operations needed much improvement. The language barrier between the public and the soldiers added to the suspicion-filled environment. I too happened to be subjected to one of such search activities when I was on leave in my village in 1980. My scooter was stopped for frisking by the army men. One young officer approached me and asked who I was. I responded by saying that I was a Captain in the army. To which he countered, 'Which b....dy Army? People's Liberation Army or Indian Army?' I quickly responded that I recognized only one army in India i.e. the Indian Army. I asked him whether he recognized any other army. That put a stop to the argument. Later I realised that it is not wise to argue with a soldier with a gun during such activities. The VBIGs (Valley-Based Insurgent Groups), as they were known by the army, employed urban guerrilla tactics and carried out sophisticated bomb attacks on the Security Forces. The UG (Underground Groups) did not specifically target any family member of the armed forces. My village and surrounding areas were infested with many militants at that time.

Very soon many splinter groups such as PREPAK (People's Revolutionary Party of Kangleipak), KYKL (Kanglei Yawol Kanna Lup), KCP (Kangleipak Communist Party) and others sprang up in the state, all purportedly fighting for the cause of Manipur. The ideology of most of the groups was partly based on Chinese communism except for the UNLF (United National Liberation Front) led by the charismatic Sanayaima. In later years many more small splinter groups were created by vested interest groups and exploited for their personal and political goals. AFSPA, which was imposed in Manipur in 1980, has had a profound negative effect on the psyche of the people especially among the younger generations. Relentless operations by the army continued for a long time. Cordon and search of houses or any locality, frisking of people anywhere and anytime, presence of armed military personnel at every nook and corner of the state became a common sight throughout the state. Frequent shootings and explosions were very routine in the years between 1980 and 2005 or so. In spite of the best efforts of the forces to prevent it, killings of innocent people did take place in the state.

The debate over AFSPA in the state of Jammu and Kashmir and in the North East is a delicate one. AFSPA is required to act as an enabler for the army to operate against anti-national elements and gives wide powers to the army. The army has internal mechanisms to ensure that the power is exercised with due caution. But violations do take place, as has been observed by the Hon'ble Supreme Court in various occasions. Kishalay Bhattacharjee in his book *Blood on my Hands*, writes, 'It is a shame that one of our most decorated officers in the service is one Major Sreeram Kumar, who was awarded the Ashoka Chakra (the highest national gallantry award in peace time), and the court charged him with fake encounter. And it happened in Manipur' (Harper Collins: p. 147). The fact is that a few gallantry awardees in such areas are known to be not genuine. The author further adds, 'In Manipur everyone is grey (suspected militant or sympathiser) there. The police mainly plotted the killings, but the army was not willing to let go of the fruits of it, so they called them "Joint Operations".' The Justice Hegde Commission has also commented on the issue. Notwithstanding, I am of the view that we in the army as law abiding citizens need to educate the troops involved about the emerging scenarios. There are elements in the ecosystem who are capable of misusing the law. The argument that the forces need to kill on mere suspicion is a tricky issue given the fact that our troops from varied social backgrounds will find it difficult to accurately assess and establish 'suspicion' in the behavioural patterns of the mongoloid looking people of the North East. The debate between the supremacy of the idea of the state and of human rights is likely to continue. I am reminded of the writings of Maj. Gen. Sir James Johnstone, who was the British Political Agent in Manipur in 1886-89 in his book *My Experiences in Manipur and the Naga Hills* where he wrote after the Anglo-Manipuri War 1891, that 'we have not heard the last of Manipur'.

Starting a Family on the Banks of the Ganges

Fatehgarh is the place where I met my wife Capt. Dr Mangala. She was posted as one of the doctors in the station military hospital. We got married a year after our courtship at my village in Manipur on 18 November 1983. Our marriage took place at the Charangpat village home. The marriage was witnessed by almost the entire village folk, though only about 350 invitations had been sent out. The big courtyard of our village home was full. It was unthinkable those days to marry someone from as far as Maharashtra. My parents agreed to our inter-caste marriage. Accordingly, Mangala's parents travelled to Manipur for the marriage. My grandmother was about 76 years old when I got married. Her eyesight was very poor. She wanted to look at Mangala on her arrival on 16 November. She saw her and felt her with her hands and declared that she was a Meitei and that everybody was fooling her by telling that she was a *Mayang* (outsider). I was 27 years old then. We enjoyed the married life in Fatehgarh. Long walks on the sandy banks of the Ganges were particularly happy moments together.

One of the most important events during the period was dealing with the Sikh soldiers who deserted their Training Centre at Ramgarh in the aftermath of 'Operation Blue Star' in June 1984. The operation was launched by the Indian Army to nab militant leader Jarnail Singh

Bhindranwale and others from the Harmandir Sahib in the Golden Temple, Amritsar. On hearing about the operation, many Sikh soldiers deserted their units and tried to reach Amritsar. One such party comprising of 35 soldiers from Ramgarh in Bihar going towards Punjab was apprehended by us. Fortunately, almost all of them gave themselves up after knowing the full details of the operation and the run-up to it. These soldiers were brought to Fatehgarh. They were accommodated in a make-shift camp under my supervision. I interacted regularly with the inmates of the camp. It was a very insightful experience for me to be with them. They were later sent back to their units. I recall one Havildar Sukhvinder Singh, an inmate of the camp, who belonged to Tarn Taran district in Punjab, who shared a special rapport with me. He narrated how the Khalsa militant groups had tried to subvert the minds of young soldiers in their villages and even in their units through the Unit religious teacher. I was able to fathom the realities of existential threats to nation building efforts of our country.

Our first daughter was born on 13 July 1984. We named her Poornima since she was born on a full moon day in the army Command Hospital, Lucknow. My parents, aunt, mother-in-law, sister-in-law and brothers visited us at Fatehgarh after the birth of Poornima. The visit to Fatehgarh was the first train journey for my aunt and my mother. In fact, it was their first journey outside Manipur. They were quite excited by the four days journey accompanied by my brother Santosh. They were taken for a pilgrimage to Mathura also. Lord Krishna is worshipped by the Vaishnavite Hindus in Manipur. My younger brother Kiran stayed with us for his studies in Fatehgarh. He was admitted in the Central School there in Class 8. My mother wanted him to study outside the state since the situation in Manipur was very disturbed in those days. Besides, my village did not have good education facilities.

In the meantime, my unit had moved to Mizoram for counter-insurgency operations in south Mizoram. I joined the unit in Mizoram when my daughter was about a year old. Mangala was posted to the Army Base Hospital in Guwahati (previously spelt Gauhati). Army life, particularly in the infantry, is quite unpredictable. We had to adjust ourselves for the separation due to the call of duty, amid managing our lives with a small baby. Mangala was unfazed. She is a woman of courage. She ensured that she cared for our child, looked after Kiran and carried out her military responsibilities equally well. It goes without saying that all this was possible with a little help from friends and relatives at that time. My friend Capt. (later Colonel) I.D. Singh from Manipur kept us for over a month in his house before we managed a small house in Narangi Military station. My mother could manage to stay and help. Our second daughter was born on 12 June 1986 when I was in Lungleh (Mizoram). Mangala had gone to Manipur for the delivery of our second child. My aunt was of great help to her. Mangala's responsibilities increased with the arrival of Priyanka.

2

Unfolding of the Challenges

Mizoram: The Night of the Guerrillas

The situation in Mizoram, a union territory then, was slowly returning to normalcy from 1984 since the Mizo Accord was in the offing. The operations by the army were reduced to a considerable extent. No serious fighting with the Mizo hostiles took place due to the ongoing peace talks though we nabbed a few insurgents carrying unauthorised weapons. My experiences during the period were confusing at times. On one occasion during 1986, hordes of civilian Chakmas from Bangladesh intruded across the boundary into India from the Chittagong Hills. There were a few thousand of them in miserable condition, half-clad, women, children and men folk. I was the Company Commander in the area. I was ordered to ensure that no Chakmas entered Indian territory. It was a humanitarian crisis of a very high order. I was permitted to use force if necessary. It appears that the Government of Bangladesh took offensive military actions against the Chakmas for their alleged support to the Shanti Bahini insurgent group in the neighbouring Chittagong Hills in Bangladesh. Shanti Bahini was a militant group which operated in the Chittagong Hill Tracts of Bangladesh to fight for the rights of Chakma tribes. Chakmas are largely Buddhist people who are inhabitants of the Mizoram-Bangladesh border. Approximately, a lakh of them are still in refugee camps in Assam-Arunachal border areas. There were reports that the Shanti Bahini was supported by Indian Intelligence agencies to counter the support of Bangladesh to the insurgent groups operating against India in the North East part of the country, such as Naga, Manipuri and Assam militant outfits. It was widely believed that ULFA (United Liberation Front of Asom) of Assam, NSCN (National Socialist Council of Nagaland) of Naga Hills and few others had established massive training and other facilities in Bangladesh with active support from Pakistani intelligence agency ISI (Inter-Services Intelligence). The civilians continued to stay in the border in the open without any shelters throughout the night. They refused to return to Chittagong Hills for fear of reprisal by the Bangladesh Army. We provided them meagre amount of food which we could afford from our limited resources to the weak and children for two days. Later, we learnt that the Government of Bangladesh had agreed to take them back on some conditions. The only government agency confronting the crisis during the standoff was my unit comprising 40-odd soldiers who tackled the situation on humanitarian considerations.

On another occasion in South Mizoram, my patrol came across some uniformed body of troops who didn't look like the typical Mizo hostiles. They were aware about our patrol's presence but made no effort to contact us or avoid us. We made contact with them by approaching them with due care in a tactically sound move. I was surprised to learn that they were part of the Shanti Bahini and were on the run from the Bangladesh Army. I asked for permission from the higher authorities to apprehend them but was told not to do so by the formation Headquarters in Lunglei and that RAW (Research and Analysis Wing) will handle the situation. Our radio sets received frequent radio transmissions of Bangladesh Army regarding sporadic fighting between them and the Shanti Bahini forces.

Mizoram is a beautiful state dominated by the Lushai tribes. However, Chakmas and Lakher tribes form over 10 per cent of the population. Mizos are 100 per cent Christian while Chakmas are mostly Buddhists. I came across many stones with carved markings indicating that British exploration of the North East hill regions commenced through this countryside astride the Chittagong Hills. The Lushai Hills were annexed by the British sometime in 1890 even though contact between them had been established by 1777. Disused telephone poles of very old vintage, possibly from Chittagong towards Aizawl, were still standing in some areas. Stone carved memorials of British preachers were also found. The border with Bangladesh had no border pillars in many places. Sights of monkey heads in the markets were common for these were delicacies among some of the inhabitants. Patrolling in the area during the period was very challenging due to the nature of the jungle. High humidity, tiger leeches, mosquitoes and reptiles made living in these areas quite uncomfortable.

During my stay in an outpost named Khojaichari on the border, I had to confine myself inside a mosquito net—for bathing, eating and resting—throughout the day and night. On return from patrolling on a rainy day, it was routine to take off the clothes and pour kerosene on the body so that the tiger leeches would fall off easily from the body. In yet another occasion our patrol noticed one huge python near a tree. We indicated the location of the reptile to the locals. One of them very coolly took out a *dah* (knife) and cut off the head of the python. The villagers rewarded us with a live chicken for giving them the information about the presence of the reptile. Pythons and monkeys were a delicacy among some of the people of that area. There were no newspapers, TV, radio, electricity, or mobile phones then. Letters from home took over 15-20 days to reach us at our outpost. At times, we authorised some officers or the head clerk to read out the contents of the letter from home through the medium of the High Frequency Radio Sets with us. We survived mostly on tinned food items. Fresh vegetables were not seen ever there. My troops did catch some fish for me from nearby rivers. Nobody complained. It was part of our duty.

My impression was that the Mizos were peace loving people. Their faith in Christianity had a lot to contribute towards the peace building process in the state. The tenure in Mizoram got truncated due to signing of the Mizo Accord between the Government of India and the Mizo hostiles led by Shri Laldenga during the Prime Ministership of the late Shri Rajiv Gandhi in 1986. Shri Rajiv Gandhi was a popular figure in Mizoram those days. During the period he even visited our unit quarter guard which was along the highway, and he halted nearby for a few minutes. Laldenga was made the Chief Minister of the state. The entire Mizo National Front (MNF) which had fought the Indian Army laid down their arms and Mizoram returned to peace, long after the fighting had broken out in 1966. The Mizo insurgency saw one of the most intense fighting in post independent India. The Mizo Peace Accord is still holding to this day. Very often people ask me as to why the Mizo Accord is holding out while accords with other groups such as Naga groups tends to fail. There could be many reasons for the same. One, I can think is Mizoram has close to 77 per cent Christian Mizo population largely homogeneous in character while the Naga society has numerous tribes that have very little in common among them. The Lushais of present day Mizoram have a cohesive society even though the larger Mizo fraternity has similar social connectivity with many tribes of Chin-Kuki origin just as the Nagas have in the form of many tribes. More on the Chin-Kuki fraternity issues later in the book. We wish that the peace holds for eternity.

The Call of the Mountains: Kanchenjunga Days

Around the time of de-induction of my unit from Mizoram after the Mizo Accord was signed, I grabbed an opportunity to join the Assam Rifles' Kanchenjunga Expedition led by Maj. Gen. P.L. Kukrety (who was a Colonel during the earlier expedition to Kedarnath Dome 1980). I readily agreed to join the expedition to the most difficult climb in the world i.e. via the northeast spur of Mount Kanchenjunga from Sikkim side. I often wonder what propelled me to undertake very risky mountaineering expeditions, sky-diving trips and other such dangerous activities. I knew that I am of average physical fitness level. I was not even a qualified mountaineer from any of the recognised institutes though I had taken part in the previous expedition to the Kedarnath Dome in 1980. Perhaps it was my love for adventure and to explore the beauty of nature more and more. The rebel in me also influenced me to take such adventures I suppose. The call of the Mountains, as they say, is a powerful call. It is said that he who climbs mountains laughs at all tragedies, real or imaginary. For me I climb mountains because it makes me feel the nothingness of life. I also wanted to prove to myself that I could do it. Many years ago, the great British explorer George Mallory, who was to die in Mount Everest, was asked (by a *New York Times* journalist) why he wanted to climb it. He said 'Because it is there.'

The expedition was scheduled to commence during March 1987. My wife Mangala was apprehensive about my physical safety during the expedition. She was looking after our two small daughters all by herself at Guwahati where she was working in a Military Hospital. She was also mainly responsible for looking after all my brothers and sister indirectly. But she collected courage and told me that I could go ahead and wished me luck. Gen. Kukrety arranged my attachment orders to the Assam Rifles for the purpose of the Kanchenjunga Expedition. Therefore, I was attached to the Assam Rifles located in North Sikkim from December 1986 onwards. I left my unit in Vairengte in Mizoram for Chatten in North Sikkim from where all our planning activities were to be carried out.

I was on my way to North Sikkim in December 1986 to prepare for the lifetime opportunity to be part of the historic expedition. The famous Teesta River originates from North Sikkim. The fauna consists of ducks, snow pigeons, ravens, ibex, antelopes, mountain goats, marmots and blue sheep. The temperature in North Sikkim ranges from 5^0 to -50^0 Celsius. The oxygen level is reduced to 10 per cent of that available in the plains. And it gets to lower than 4 per cent as one climbs higher than 20,000 feet. Planning the expedition at such a short notice was a herculean task. Luckily, Gen. Kukrety had done most of the homework of obtaining permission to climb Kanchenjunga from Sikkim side which was far tougher than climbing from the Tibetan side. He had also obtained reasonable amount of funding from the MHA (Ministry of Home Affairs). We had to concentrate on procurement of clothing and equipment from Nepal and other sources. Packing, accounting of the items were also major tasks on hand. Maj. Stobdan, an experienced mountaineer, was the deputy leader. A party comprising myself, Mr Phu Dorjee of Everest fame under the leadership of Maj. Stobdan, proceeded to procure mountaineering equipment from the Kathmandu market in Nepal. Under the guidance of Gen. Kukrety and Maj. Stobdan we were able to complete procurement of the equipment, and returned to North Sikkim by January 1987.

The expedition members were mostly from the Assam Rifles. Two climbers were from the Sonam Gyatso Mountaineering Institute, Gangtok (SGMI) which was under the MHA. Two female mountaineers from Sikkim were also members. The expedition was the second in the history of mountaineering to attempt to scale the peak from Sikkim side. The first being an Indian Army expedition of 1977 led by Col N. Kumar. With an altitude of 28,208 feet above sea level, Kanchenjunga is the third highest mountain in the world after Mount Everest and K2. Among thc people of Sikkim and the surrounding areas, Kanchenjunga deity is revered along with various manifestations of the Buddha. Kanchenjunga literally means 'five treasures of the snow'. The people of Sikkim believe that the all-powerful God on the peak of the mountain controls their destinies.

As per historical records, one W.W. Graham in 1883 led the first expedition to the Kanchenjunga massif. The expedition claimed to have climbed upto a height of 24,000 ft or so. Many more expeditions followed thereafter. In 1955, a British expedition led by Charles Evans made the summit from the north-west face on the Nepal side. But none had climbed the peak from the Sikkim side. It was on 30 May 1977 that an Indian Army expedition team led by Col N. Kumar was able to summit the peak. Major Premchand and Naik N.D. Sherpa were summiteers. Our expedition also aimed to put the lady climbers of Sikkim on top of the revered mountain. Members included Everesters Phu Dorji and Nima Sherpa. The expedition also aimed to scale the peak at a mere fraction of the cost of what the Indian Army incurred in 1977 thereby making this sport affordable to more and more people across the country. We had assured the Government of Sikkim that no climber would put their foot on the peak. Kanchenjunga is revered as a Deity by the Sikkimese.

We set course towards the Green Lake (Base Camp) along the Zemu glacier from the mouth of the glacier on 9 March 1987. We had planned to establish the advance base camp by 24 March, thereafter attempt to establish Camp 2 on the most treacherous northeast spur well before 1 May. Such a plan required the blessings of the weather god and Mother Nature. With Himalayan effort by Phu Dorji and Nima of SGMI and a few other climbers we succeeded in establishing the camp on the NE spur by 10 May. The Base Camp and the Advanced Base camps were established as per our plan. The beauty of the alpine lakes, and of the surrounding peaks of Mount Simvo, Mount Siniolochu and Tent Peak were unforgettable sights. The serene beauty of the mountains was briefly disturbed by the loud noises of the massive avalanches of ice and rocks from the surrounding mountain faces. Some of us had medical issues including myself. I was unable to proceed beyond a height of 23,400 ft due to high altitude sickness and was advised to fall back to the advance base camp by Capt. Gopal, the doctor.

The most challenging climb of the entire expedition in terms of technical climb was overcoming the northeast spur and setting up of Camp 3. The climb involved cutting through the vertical ice walls and cutting vertical ice tunnels. Climbing through the ice overhangs by hacking each and every step, hammering pitons, and putting ropes took considerable time and effort. I was confident that once the northeast spur was overcome, it would be mostly snow fields albeit with very steep ice faces. It was a matter of stamina, the body and the will power thereafter. We had to establish three more camps along the spur to be able to attempt the summit. Maj. Stobdan was on the lead after Camp 3. He ensured that Camp 5 was in place by 20 May. There were some hiccups about the logistics backup due to weather conditions.

The first attempt to summit was made by Phu Dorji and party on 23 May. A big tragedy occurred on this day when we lost radio contact with the team at about 1 p.m. Efforts to link up with the team failed due to hostile weather conditions. Search operations did not result in any good news. Days later we presumed that the party of five had possibly been swept away from near the summit by strong blizzards in the area or slipped down into the crevasses while climbing roped up. It was a big loss of morale for all of us. We tried our best to locate the lost climbers but to no avail. Phu Dorji was a world famous Everester. A great mountaineer had been lost. A second attempt was made on

31 May by another party led by Naik Puran Singh. On 31 May, the expedition succeeded in summiting the peak a few metres below as promised to the Sikkim Government. One of the oxygen cylinders which had been used by Phu Dorji himself was found 50 metres below the summit. The expedition was a grand success. Greetings from the Prime Minister reached us on our return to the base camp. I was the last man to leave the Base Camp on 5 June 1987.

In the Jaws of Death: Inside a Yawning Crevasse on Zemu Glacier

I had a very narrow escape when I slipped inside a crevasse while negotiating an ice face during the return from the Advance Base Camp. I was on the safety rope with my carabiner in place. I was a few metres behind the party of four and on a turn of the route when I slipped and fell inside a yawning crevasse. The rest of the party could not hear my call for help when I slipped into the mouth of a gaping crevasse at an altitude of about 16,000 feet. I was dangling about 20 feet inside the crevasse with the possibility of the ice pitons further giving way due to the melting ice. My energy level was at the minimum. I could not pull myself up any further to be out of the crevasse. The more I struggled to come up, the more the rope sagged and went down further. I could only pray to God. When everything seemed hopeless, it was a miracle that Havildar Surjeet of my party returned by the same route looking for me since he did not see me for a while. He heard my call for help and called the other three also to rescue me. It was getting dark and rescue became more difficult. With the grace of God, the three mountaineer colleagues rescued me from the mouth of the bottomless yawning crevasse. I overheard the conversation of helplessness of the three as they struggled to get me out. I prayed to God, 'Please don't let them go away without getting me out of the place.' God heard my prayers. They struggled for half an hour to get me out of the jaws of death. In the bargain, Surjeet suffered cuts in his hands through the gloves. This half hour struggle for life inside the crevasse taught me of the 'Nothingness of Life'.

Reinhold Messner, one of the greatest mountaineers of all time, once remarked 'Mountains are not fair or unfair, they are just dangerous'. He also said: 'Without the possibility of death, adventure is not possible.' This particular incident was the closest I got to near-death situation in the mountains. It was carelessness on my part to be the last man out and yet not keeping up with the rest of my party. Such experiences teach us of the smallness of humans against nature and of the nothingness of life. We returned to the mouth of Zemu Glacier on 7 June after 95 days of continuously being on the glaciers and mountains. On our return, General Sundarji's 'Operation Falcon' was in progress in these Himalayan heights to counter the Chinese aggressive posturing on the LAC (Line of Actual Control) all along the Tibet Border.

This was my second expedition. The lessons learnt from this expedition remained throughout my life. Had it not been for the loss of precious lives, the expedition would have gone down in the annals of mountaineering as one of the most daring ones. The overall cost of our expedition which took place more than 10 years after the army expedition 1977 was merely a fraction of the cost of the latter expedition. We were not successful in putting the Sikkimese ladies Ms Dikila and Ms Yangtse on the summit. They were with my team till Camp 3. They were very spirited and tough climbers.

The entire team was called for an audience with the Prime Minister at his official residence. Ms Dikila and Ms Yangtse were also in attendance. It was my pleasure to have briefed Prime Minister Rajiv Gandhi on the expedition along with Maj. Stobdan. Unfortunately for Gen. Kukrety, he was not there to meet the Prime Minister due to a raid on an Assam Rifles post at Oinam (Manipur) on 9 July 1987 by the NSCN(IM) (National Socialist Council of Nagaland-

Isak-Muivah) in which they had killed nine soldiers and carried away 150 guns and 1,25,000 rounds of ammunition from the post. It was a huge setback for Assam Rifles. Gen. Kukrety opted to stay back at Imphal to attend to the urgent operations at hand. The audience with the Prime Minister was a great opportunity for the mountaineers. Shri Rajiv Gandhi appreciated the team members and narrated to us about his love of the mountains too.

Oinam: Operation Bluebird 1987

On completion of the Kanchenjunga expedition in August 1987, I was ordered to assume the duties as GSO3 (General Staff Officer) in the Headquarters of Maj. Gen. Kukrety who was the General Officer Commanding (GOC) 'M Sector' and the Inspector General Assam Rifles (North) (IGAR N) located at Imphal. I started to write a draft of a book on the expedition to the Kanchenjunga 1987 as directed by the General. I could put together bits and pieces of the events of the expedition in my spare time. The challenges of winding up the expedition which included detailed accounting, disposal of the mountaineering equipment and stores, recognition to the achievers, and other issues were in progress too. I was able to hand over a draft of the book before the General superannuated in early 1989. However, it was learnt years later from Mrs Kukrety that the General had handed over the draft to someone for publication along with the photographs. After his sad and sudden demise in October 1989 we have not been able to trace the draft till date. The book was aimed to sensitise the people of the fact that mountaineering as a sport should be available to any mountain lover and not only be for the elite.

M Sector was in operational control of all army and Assam Rifles units and formations engaged in Counter-Insurgency operations in Manipur. I returned to Manipur when 'Operation Bluebird' against the NSCN was in progress during July 1987. This operation was launched by Gen. Kukrety with an aim to apprehend the culprits of the NSCN as also to recover the arms and ammunition which were taken away from an Assam Rifles posts after a raid on 9 July 1987 in which nine soldiers were killed. All the human rights activists were up in arms against the Assam Rifles. Local civilians were also very agitated by the hardship and alleged torture by the Assam Rifles. There were many allegations of human rights violations by the forces. Events leading to and the 'Operation Bluebird' itself was a very unfortunate period in the history of Manipur. While the 9 July raid marked a major push by NSCN leader Thuingaleng Muivah to consolidate his views on the Naga cause, it was also a watershed for the army and the Assam Rifles which had been taking many casualties. The human rights issue became a very important matter during the period. National and international human rights activists were in full swing in the state. At the end of the operation 80 per cent of the arms and ammunition were recovered from the nearby jungles. In all 27 civilians were also killed in the crossfire. The Assam Rifles used strong arm tactics to elicit information from the civilians in the area. The NSCN had used many of the civilians as porters to carry the warlike stores after the raid and killings of the Assam Rifles personnel. Use of innocent civilians forcefully to take part in the nefarious designs of the NSCN was also responsible for the entire episode. It was imperative that the writ of the state was ensured by the Assam Rifles in the area. In sum, I would like to forget the entire episode as a very unfortunate chapter in the history of Manipur.

It appeared to me that Gen. Kukrety was able to ascertain the identities of some high ranking individuals in the Government of that time who were directly or indirectly helping the cause of the banned militant outfit. During the visit of Mr R. Venkataraman, President of India, to Manipur later in 1987, Gen. Kukrety found an opportunity to make a detailed presentation to the President regarding the situation in Manipur. The venue was at Raj Bhawan, Imphal. Gen. Krishna Rao was the Governor. I was asked to operate the slide show during the presentation. Gen. Kukrety commenced his presentation by addressing the President as Supreme Commander of the Armed Forces of India. He also told the President that he was addressing him as the Supreme Commander since the President, as the Head of State may not like Gen. Kukrety's exposé on the involvement of some figures in the state in the anti-India activities. It was then that the Governor asked me to leave the three of them alone in the room! I left the room and later Gen. Kukrety shared with me what had transpired between them.

In 1975 the Naga separatist Naga National Council had signed the Shillong Accord and gave up arms. Those who did not agree with the accord formed NSCN, and it was led by Isaac Chisi Swu, T. Muivah and S.S. Khaplang. On 30 April 1988, the NSCN further split into two factions. NSCN(K) was led by a Myanmarese Naga, Mr Khaplang and NSCN(IM) led by Mr T. Muivah of Manipur. The outfit runs almost a parallel government in their sphere of influence. Extortions and recruitments by them is considered normal in these areas.

During October 1988, we received a letter possibly from one of the workers or sympathisers of the NSCN with a threat to the life of Gen. Kukrety. But the General was no pushover. He told me that we will go to Ukhrul by road (90 km) after informing the NSCN! I politely reminded him that it was no use being foolhardy by telling the NSCN, thus inviting trouble for ourselves. We decided to go driving to Ukhrul incognito in a Maruti Gypsy. We carried some arms and ammunitions and the General was driving all the way to Samsai (Ukhrul). Such was his courage. He predicted in 1987 in confidence to me that one day Manipur state boundary as we knew it then was likely to be seriously threatened due to the combined forces of politics and the gun. I could not believe his prediction then. He narrated to me why he considered that the existence of Manipur as we knew was under severe threat from the NSCN (IM) in particular. He was of the opinion that while the valley based militant groups were for a united fight against the Indian forces, the NSCN militants were fighting for a Christian State comprising huge areas of Manipur, Assam and Arunachal Pradesh. He also opined that while the Manipuris are extremely divided in the approach, the Nagas were united in the aim to integrate all Naga dominated areas one day. He once mentioned to me that he would do whatever it takes to save Manipur as a state entity as it existed at the time. I think here was a man who saw the future!

In the Lap of Kangla

By this time Mangala had completed her five-year Short Service Commission in the Army Medical Corps. She shifted to Imphal from Guwahati where her last assignment was. We were allotted a newly-constructed house inside the Kangla Complex. Kangla is a very important historical site of the Manipuris. The British occupied the Kangla Fort after defeating the Manipuri army in the Battle of Khongjom in 1891. This dilapidated old fort has been under the occupation of Assam Rifles after the British left Manipur. The Manipuris had been legitimately demanding vacation of the Kangla Fort by the Assam Rifles for a long time. The fort is on an area of 237.62 acres. Most structures inside were destroyed by the British to remove any semblance of Manipuri power and make way for their facilities. After the vacation of the fort by the Assam Rifles on 20 November 2004, the Kangla Fort Board under the Chairmanship of the Chief Minister manages this historical site. The place is now one of the major tourist attractions. The house allotted to us on arrival inside the Kangla complex was a comfortable two-bedroom cottage. Along with my wife and two small daughters, we were able to keep my siblings Anita and Kiran for their studies at Imphal. My parents and relatives visited us frequently. It was a great privilege for me to have the opportunity to be residing inside the historical Kangla Fort complex. The complex was also used as the Headquarters of Field Marshal William Slim, Commander 14th Army of the Allied Forces during the Burma Campaign in 1942-44. Gen. Kukrety became a family friend of ours. He was not only from the Rajput Regiment but I had also been on two mountaineering expeditions with him. The General and his family were invited to my village home in Charangpat village for meals. Sometimes he stayed over for a day or two in the village home with his entire family. Once he came with actor Naseeruddin Shah and his wife Ratna Pathak, singer Asha Bhonsle and composer R.D. Burman to our house in the village for a Manipuri feast.

I served with him during very trying circumstances. It was during his tenure that there was a mutiny-like situation in one or two units of Assam Rifles over the issue of the Fifth Pay Commission. The General would visit any unit wherever trouble was brewing. One Commanding Officer of one such unit advised him to abort the visit fearing lawlessness on the part of the jawans. I heard him speaking to the CO telling him to call the Second in Command. Thereafter, the General ordered the CO to hand over the command of the unit to the Second in Command, and the CO was removed from command for showing lack of command and control. He did visit the unit to calm down the emotions of the troops in spite of the simmering discontent and possibility of an ugly situation erupting there.

The powers and responsibilities of M Sector were equivalent to that of an Army Divisional Headquarters. However, the organisational structure, staff manning and Standard Operating Procedures left much to be desired. It seemed to me that the Headquarters was functioning due to the sheer force of the persona of Gen. Kukrety. He was either loved very much by the civilian populace who knew him or he was hated by many who formed their impression based on newspaper reports. There were no TV channels or social media facilities at that time. Some of the newspaper reports were mostly influenced by the gun or cash in many instances. AFSPA was a very sensitive issue in Manipur and it continues to be so till date. The Supreme Court of India and the Justice Hegde Committee have adversely commented on the misuse of the AFSPA by the state apparatus.

Legends of Manipur

I had the honour of meeting many prominent personalities of the state during my tenure in Manipur. One of my most memorable associations was with *Pabung* (a respectful term signifying father-like figure or father) the late Captain M.K. Priyobarta Singh, the younger brother of the last King of Manipur Maharaja Budhchandra. He is the second son of Sir Churachand Singh KCSI, CBE (British Empire honours), Maharaja of Manipur. He was also the first Manipuri Chief Minister of the independent kingdom of Manipur in 1946 in the Interim Council. He was commissioned in 2 Assam Regiment during the World War II. His contribution as the President of Manipur Cultural Integration Conference until his death on 29 October 2005 was remarkable. I learnt many things about my state Manipur from him, its history, its people and his vision for the future of Manipur. I dare say that if there was one person who knew the pulse of the people of Manipur in the hills and plains, it was him. He was much against the British policy of artificial isolation between the hills and plains people. At times he lamented his helplessness over this state of affairs. Once he took me to the house of his old friend Major Bob Khathing of the same Regiment who himself was a legend in the hills of Manipur. Major Khathing, a Tangkhul from Ukhrul District, had served as a Minister in the first cabinet of independent Manipur state in 1948, before merger to the Indian Union. He had been India's Ambassador to Burma. He is known for his contribution to consolidate India's position in Tawang in present day Arunachal Pradesh. I consider it most lucky to have been associated with him and picked up some of his wisdom. I had the honour to be invited to the book launch by John Khathing on his late father at his Mantri Pukhri home on 26 February 2020.

Mr Tompok Singh, an ex-MP and the Deputy Chief Minister of Manipur, was another visionary with whom I had the privilege to be associated. Gen. Kukrety was very fond of Mr Tompok due to his nationalistic thoughts and actions. We lost a visionary in Mr Tompok who died in an air crash during 1992 in Manipur. Both of them saw the danger which was being posed to the existence of Manipur by forces inimical to the idea of Manipur. Both worked towards a strong India and believed in the greatness of Manipur. Nongmaithem Pahari, a legendary singer whose song *Sana Leibak Manipur* enchanted the youth of Manipur in the 1970s, shared his vision of Manipur with me. He had been one of the co-founders of an insurgent group in Manipur. Lt Col H. Bhuban Singh was the first Commissioned Officer from Manipur in the Indian Army. After his retirement in 1977, he was re-employed in the local administration in various capacities. He was also the first from the state to reach the rank of a Lt Col in the army. He later became the Chairman of the State Public Service Commission and the President of state BJP (Bharatiya Janata Party). I have very fond memories of my interaction with him. His simple life, impeccable integrity and disciplined life is worth emulating by us.

Gen. Kukrety was due to retire in February 1989. He confided in me that he intended to work for preservation of the integrity of Manipur and its people after his retirement. I mocked at him on how he could do anything for Manipur after retirement. He believed that he was on the hit list of the NSCN (IM) those days. Nonetheless, he insisted that I arrange a place for him to stay near Imphal after his retirement, which I did. He narrated to me why he chose to stay on in Manipur after his retirement. He told me that there were a few people in Manipur who could save the state from breaking up and protect the people. He was going to be one of them. We purchased the land from Shri Kebisana, the younger son of the last Maharaja of Manipur Shri Budhachandra Singh who signed the merger agreement between the Government of India and the Kingdom of Manipur in 1949. After the privy purse was discontinued, there was hardly any income for the young sons of the king. They had to sell bits and pieces of their property in order to survive and protect their status as well. There were multiple legal issues involved after the death of the last Maharaja as regards the titles and properties of his heirs. A small two-room hut was constructed near the existing palace. Unfortunately, fate snatched the life of Gen. Kukrety in October 1989 due to cardiac arrest in Delhi.

As I write my memoirs today, it is unbelievable that the words of *Pabung* Priyobarta, Gen. Kukrety and *Pabung* Tompok ring in my ears. They saw the future. They saw what was being plotted in the jungles and hills by various groups to challenge the Indian state with active support from elements close to the governments in countries like China, Pakistan and Bangladesh at that time. The future trajectory of security of North Eastern states of Manipur, Assam and Arunachal Pradesh was seriously impacted by the activities of NSCN.

My 18-month tenure in Manipur was an eye-opener for me with regard to the prevailing ethnic tensions and socio-political issues being faced by the state. Most valley dwellers, in particular the Meiteis, were quite happy in their own world of enough food to eat, drink, lost in innumerable festivals which included Hindu festivals, native festivals of all ethnic groups, *Leikai* (locality) politics etc. The common people in the valley were quite oblivious to the fast changing socio-political canvass in the hills and beyond. I suspect that the political leaders simply swept these issues under the carpet wishing and hoping that someday the issues would disappear on their own. The changes since the early 1960s were being spearheaded with the ideology of sub-nationalism, ethnic identity and its manifestations were in terms of armed rebellions. There was no common platform where interactions between the groups could progress on a positive note. Nor did any of the leaders at the time try to make one in a sincere manner.

Militancy and armed rebellions against the government had engulfed the entire state including the valley areas by the early 1980s. No one asked questions, no one cared. Those who asked uncomfortable questions were eliminated with ease. It is said that quite a few of the stakeholders like some of the leaders, politicians, businessmen, local clubs, the police, the contractors and bureaucrats benefitted from the state of affairs. Common people seemed to not care at all. They chose to suffer in silence. Dozens of militant groups sprang up overnight in the state during 1980-90s to claim legitimacy as true fighters for the state. I knew the time was not far that the cross purposes of the narrative of the various ethnic groups would confront one another on some pretext in the not so distant future. Sad but true.

The situation in Manipur as I saw during the period warranted immediate course correction on the part of the political and other social elderly leaders. In Manipuri there is one word to describe the total condescending attitude i.e. *He Ra*, meaning I care a damn. Most couldn't see what was coming. Those who knew what was likely to come didn't care. Looking back now, I regret to remark that our leaders of that time failed the future generations by their omission to address the ethnic tensions that were building up over a period of time soon after merger of Manipur with the Indian Union. The 20-year transition period from the time of grant of statehood in 1972 to the 1990s has been a rather tragic period. The management of change in the society had not only failed, but a regressive culture in many spheres was noticeable.

Wellington Days

I proceeded to the prestigious Defence Services Staff College at Wellington, Nilgiri Hills, in May 1989 after clearing a stiff competitive examination among the officers. Two army officers from Manipur had in the past qualified and attended the course. They were Col I.S. Chanam of Corps of Signals and Col A.G.S. Koijam of the Gorkha Rifles. Wellington is a small hill station with most beautiful natural surroundings. The course was a year-long studies on military matters. It was considered a stepping stone to higher ranks in the army. Training at the Staff College was one of the most enjoyable and learning experience in my early years in the army. The Commandant Lt Gen. Brar was a flamboyant General Officer with a great sense of wit, humour and wisdom. So were the other Directing Staff. They ensured that we learnt the mechanics and the art of war at Brigade and Divisional levels so that we could become better staff officers at these Headquarters. The curricula included every detail of staff work in any situation we were likely to face in the future wars, including calculation of number of mules required to transport rations for a body of thousands of troops in the mountains! We sharpened our operational skills, logistics skills, man management skills in the 11 months that we spent in Wellington. The pristine environment of Wellington garrison was simply magnificent. The entire family enjoyed the stay there. My father and brother Santosh visited us during our stay there. Mangala had temporarily left her job as a Doctor in the Assam Rifles hospital to be together with me. A few months before my course was to terminate, she received orders to re-join duty and accordingly Mangala, Poornima and Priyanka left for Manipur in March 1990.

One of the most interesting activities in Wellington was the Guest Lectures by eminent personalities of India. During one of such lectures on the insurgency in North East India, a Lieutenant General (name withheld) made a very uncomfortable remark about counter-insurgency operations in Nagaland and Manipur. He made a remark to suggest that the youth of the area had to be made soft and spoilt with free supplies of liquor so that they would become weak and insurgent activity would wane or words to that effect! I took offence to the remark and walked up to the Commandant with my objections during the customary tea in the Chanakya Lawns. I was introduced to the visiting General and he took his words back. The damage was already done. The 250-odd army officers attending the course would have gone away with a wrong perception of things there. The General obviously did not expect a Manipuri officer to be present in the elite crowd. Many of my colleagues in the course from the army conveyed to me that such a remark by a senior officer was uncalled for.

The tri-service environment of the course made the learning experience a truly memorable one. Officers from the Army, Navy and Air Force trained together in the art and science of war. There were many officers from other friendly foreign countries as well. I took keen interest in the war games in particular. War games are nothing but two-sided exercises under a controlled war or battle situation. In retrospect, I am of the view that the training at the Defence Services Staff College could have been far more beneficial had the curriculum been focused on the relevant current and future nature of warfare. The syllabi was mostly on staff duties in various types of warfare at the formation levels. It has been my observation over a period of time that many staff officers actually delay, confuse and misinterpret a commander's intent to the troops. The Directing Staff body were professionals of very high repute whose expertise could have been better utilized to impart very useful training to the student officers. The issue of Jointness (cross-service cooperation) between the services could have been one of the Key Result Areas (KRA) of the College.

Unfolding of Proxy War in J&K

The operation to annexe Jammu and Kashmir, codenamed 'Operation Topac', was said to have been formulated by the Pakistani ISI (Inter-Services Intelligence) in 1988. The period 1987-89, which marked the initial phases of the insurgency, was characterised by sporadic bomb blasts, killings and kidnappings, targeting the Kashmiri Pandits etc. Radical Islam had taken roots in the Kashmir valley. After the Officers course ended in 1990, I joined my unit deployed at Nowshera area on the LoC in J&K. The situation in the state had gone from bad to worse during 1989-90. Reasons for such a state of affairs were many. Governance had practically failed for many years; withdrawal of Soviets from Afghanistan left many fighters unemployed in Pakistan; General Zia-ul-Haq was on the way to radicalise the Pakistani army to snatch Kashmir from India by waging a kind of jehad. Massive infiltration of militants sponsored by the Pakistani army and the ISI made the situation grim in J&K. The deployment of the Indian Army at the LoC that time fell short of countering the infiltration of that magnitude. Soon, we had to re-orient our deployment to ensure robust counter-infiltration measures, tactically, technologically, physically and mentally too. The entire focus was to nab or neutralise the militants. There were many gaps and passes in the terrain under the unit's area of responsibility where we had to deploy extra forces by way of laying of multiple ambushes, patrols and other forms of surveillance.

As a Company Commander I had to be out in ambush operations to nab the militants almost every third day physically in the ravines and jungles. Firefights broke out at the drop of a hat between us and the Pakistani troops. Once I was speaking to my Commanding Officer Col Thomas Mathew on telephone during one of these incidents. Engrossed in the conversation with the CO, I forgot that I was in the direct line of fire of the enemy automatic gun just a few hundred metres away. I was lucky to have escaped being a casualty when one enemy stray bullet hit the telephone hand set which I was holding in my hand! I still preserve the bullet with me. It was a miraculous save by the grace of God. Soon the Armed Forces Special Powers Act (AFSPA) was also enforced in the area of a 5-km belt from the LoC. Many intelligence reports were received regarding possible infiltration by the militants in my area of responsibility. I was sure that our counter-infiltration grid was watertight and made their infiltration attempts futile.

We also came to know that many young people from the foothills of Pir Panjal like Poonch, Mendhar and Surankote had gone over to Pakistan Occupied Kashmir (POK) to receive training in guerrilla warfare and to wage war against India. Nowshera area was Hindu-dominated and relatively free from such activities during the period. The situation in the valley in the north of Pir Panjal was more serious. However, Poonch and Rajouri districts in the south were also affected by the so-called Kashmiri freedom movement upto 1991. By 1992, the Pir Panjal Ranges and the foothills of Mendhar, Surankote had become hotbeds of militancy. Most of the small-scale operations in counter-insurgency and counter-infiltration are directly carried out at the Company level. Ambushes, patrolling, sniping, search and destroy missions, are part of the tasks at this level. A fair amount of our activities revolved around WHAM (Winning the Hearts and Minds) among the local populace, without whose support the jehadi militants were fish without water.

I had more share of enjoyment as a Company Commander of a Rifle Company on the LoC. I sighted bears, pythons, wild boars and other wild animals in the hills and the forests. The Pakistani troops deployed in the proximity never objected to our hunting wild boars. Our troops do not like non-vegetarian food, certainly not pork. I had encounters with pythons, leopards and giant monitor lizards very often while moving from one place to another along the LoC. Even leopards were present in good numbers. I saw the remains of the Old Mughal Route in the area where stone carved elephants were constructed. The Mughals took the route through the area while travelling from Delhi to Srinagar via Bhimber (now in POK), Chingus, Bafliaz (named after the horse of Alexander), Pir ki Gali on the Pir Panjal and then to Srinagar. The location of my sub-unit was astride the Old Mughal Route and the stone-carved statues of elephants were erected as a mark of welcome to the emperors by the local 'Subedars'. The famous fort of Chingus which was nearby was the resting place of the Mughals while marching from Delhi to Srinagar Valley. These were mostly in ruins then.

The minefields laid during the wars of 1948, 1965, 1971 have also prevented the Mughal route along the LoC from being maintained. My tenure as the Company Commander of a Rifle Company under the direct observation and fire of the Pakistani army gave me ample opportunities to learn and practice the art of fighting small-scale conflicts at sub-unit levels. I enjoyed dealing with my troops during the command assignment. Much of the fun was to be alongside the troops, playing, shooting, facing and retaliating to the enemy fire in equal measure, laughing and joking together. One night, I received frantic calls from a satellite post informing me that one jawan had been hurt by a bear. I arranged his evacuation and later found out that the bear had been sleeping near the jawan, and the jawan had unknowingly passed on his *bidi* (hand-rolled cigarette) to the bear thinking that it was his friend near him! On seeing the lit bidi, the bear panicked and clawed the jawan's face!

The Simla Agreement of 1972 prohibited construction of new defensive positions or structures in the FDLs (Forward Defensive Localities). Hence, the defensive positions on both sides of the LoC had been considerably denuded. In the light of emerging militancy and infiltration of the militants from the POK side we had to take measures to thwart these attempts. Local resources of stones, rocks, fallen trees had to be commandeered to strengthen the positions. Ceasefire violations, infiltration attempts and frequent clashes had become routine. A little less than 20 years later I visited the same Company deployment in 2011, now in the capacity as the General Officer Commanding the Division in that sector. It was a nostalgic experience to look at the fighting and living bunkers which were made during my tenure as the Company Commander. Much had changed. The defences were far more robust to withstand enemy artillery shells.

After a year in the unit in Nowshera, I was posted on compassionate grounds to Brigade Headquarters at Leimakhong in Manipur. I had applied for a posting near my home to look after my mother who was ailing, besides the unit was also to be de-inducted to a peace location at Faizabad (UP). It was another big blow our family and me when my mother passed away in 1992. Mangala had done all she could do in my absence. On being detected with cancer, she took my mother to Guwahati in a taxi travelling throughout the night from Imphal. In those days, it was unthinkable to travel alone in a taxi to Guwahati. She made sure that my mother got the best treatment available that time. She was the real strength in the family. She was supportive of all our endeavours. All of us became worthy children of our parents only due to their blessings.

The Calling in Manipur Once Again: Ethnic Unrest

Once again I was back in the insurgency-infested area of Manipur, in another capacity in 1991. Although incidences of militant attacks on the Security Forces (SF) were much reduced in the valley area, activities such as extortion and kidnappings had increased significantly. Many of the sympathisers or active cadres had infiltrated into the government services and in contract work. This tenure also saw the emergence of militant Kuki sub-nationalism. It was manifested in the form of armed groups mobilizing to resist the NSCN (IM) inroads into what Kukis claimed as their sphere of interest such as land, identity, and illegal taxation. The erstwhile insurgencies were slowly getting morphed into identity wars to protect their respective group interests. The NSCN (IM) also established significant influence in the valley areas with active collaboration from one or two valley-based groups. The tenure during 1991-93 was marked by very unfortunate episodes of Naga and Kuki killings, Meitei-Pangal (Manipuri Muslims) communal clashes, major NSCN ambushes on the Security Forces near Machi village in Chandel district and Lokchao near Moreh. Historically, Naga-Kuki clashes are not new in Manipur.

The Naga tribes and the Kuki tribes dominate the northern and southern part of the Manipur Hills respectively. The Kukis belong to the Chin-Kuki group while the Nagas are believed to be of Tibeto-Burman stock of the Mongoloid race. The Kukis have affiliation of customs and dialect with southern neighbour Mizoram and Chin Hills of present day Myanmar, while the Naga groups have some similarity and affiliation with the Nagas of Nagaland and Myanmar. The Meitei Hindus who are mainly concentrated in the central valley area have had a chequered relationship with some of their tribal brethren in the hills. I experienced this psychological divide during 1977 when I tried to take my colleagues, the late Thomas Khangra, a Tangkhul, and Sitlou, a Kuki, from our state for a break in Delhi. All three of us came to Delhi during the winter break from the Indian Military Academy Dehradun (IMA) where we were undergoing training to be commissioned in the army. Christmas celebrations were planned in all the three State Bhawans (Manipur, Nagaland and Mizoram). Even though all three of us were from Manipur, Thomas preferred to go to Nagaland House while Sitlou preferred the Mizoram House for the celebrations. We were very good friends though I was a term senior to both of them. I told them that we will all go together wherever we go and that we would not get separated whatever the case be. We later decided that we will go to the Tibetan colony for the evening after a brief visit to all the State Bhawans. They agreed. This episode explains the complexity of thoughts in the ethnic rivalry even today. Both Thomas and Sitlou told me that it was not so much about the divide that I was imagining, but of the joy of celebrating Christmas among the Christians! Thomas had a tragic end in a train accident after his commission in 1979.

During the colonial period, frequent fights between the two communities broke out over land rights, or hunting rights. Claims, counterclaims over land citing 'time immemorial' unrecorded (?) 'historical evidences' mostly constructed by respective tribes, and political jockeying for land rights in the hills are becoming more and more shrill and intense day by day. Their respective agendas and the narratives are propagated through the barrel of the gun where required. Demand for Nagalim (Naga Country) and Kuki Land overlap in the areas of Chandel District, Senapati District, Kangpokpi District, and Tengnoupal districts of the state. Nagas claimed that Kukis were migrating tribes and were relocated in these areas from the Chin Hills of Burma by the British and have been traditionally paying tax to the Nagas. The Nagas in Manipur also claim that the indigenous groups in Manipur are Meiteis and the Nagas only. Col James Johnstone in his book *My Experiences in Manipur and the Naga Hills* (1875) has written:

> Kukis were first heard of as Kukis, in Manipur, between 1830 and 1840; though tribes of the same race had long been subject to the Rajah of Manipur. The new immigrants... soon poured into the hill tracts of Manipur in such numbers, as to drive away many of the older inhabitants.... the Rajah Nar Singh gladly handed over the management of the new arrivals to Colonel McCulloch.... He

> settled them down in different places where they would be useful on exposed frontiers...

The Kukis on the other hand claim that they are original inhabitants of most of the hill areas in the state. They also claim that many of the so-called Nagas are in fact Old Kukis, i.e. Anals, Marings, Moyon, Chiru and so on. T.C. Hodson, erstwhile British Political agent in the kingdom of Manipur during 1908, in his book *The Naga Tribes of Manipur* has detailed the laws and customs, religion, folk tales and background history of each of the Naga tribes separately. His treatise includes the Tangkhul, Mao Marams, Kabuis, and a few other smaller tribes such as Marings, Khoiraos and Chirus which indicate no common ancestry as such. Each Naga or Kuki tribe has varying customs and laws. The rivalry between the Naga tribes and Kuki tribes has existed for centuries. There being no written record of early history of any kind except for the *Cheitharol Kumbaba* (Historical chronicles of the Kings of Manipur) where mention is made of the tribes in a scanty manner, it is well-nigh impossible to trace the validity of the claims and counter-claims of the parties at this stage. The rulers of Manipur, the Meitei Kings, had also taken advantage of the rivalry to enforce rule of law in the hill areas of the kingdom from time to time.

But the immediate cause for the 1992-93 clashes was primarily over land, and control of the smuggling hub of Moreh township at the border between India and Myanmar. Kukis claimed that they lost 1,157 people including children and womenfolk. During the '80s, Naga insurgents had been collecting taxes from the Kuki settlers in their area and even in the Moreh area where Kukis are in the majority. In 1990, the Kukis refused to pay taxes. This was seen as a direct challenge to the authority of NSCN in the area and was taken as an intent to take on the NSCN with force. The multi-crore smuggling business of drugs, contraband goods, arms and other goods originating from neighbouring Myanmar, was also a major bone of contention between the two. In an interview with *North East Sun*, NSCN leader Thuingaleng Muivah says that 'The intention of KNA (Kuki National Army) was to chase Nagas from Moreh and the foothills near Imphal. This is the crux of the problem. Our men went and attacked the Kuki army camp because they had started forcing Nagas to leave the area. Once the foreigners started chasing the indigenous people we had to take action.' Kukis, however, blame the Nagas for their intent to create a Naga nation for Christians which includes Kuki areas. Kukis also accused the Nagas of continuous domination and harassment since Independence of India in 1947.

It was difficult for the army troops to prevent the ethnic conflict. Our troops organised various inter-community meetings in the interior villages to bring about peace. During one of my vehicle-mounted patrols in the affected area, I myself had to use force to separate the two warring parties near Tengnoupal village. My party evacuated many of the seriously injured belonging to both the warring factions to the hospitals in Imphal. My thoughts go back to the prophecies of the likes of Pabung Priyobarta, Sri Tompok Singh and Gen. Kukrety who had foreseen this kind of future in Manipur. During the late 1980s, NSCN had circulated some controversial books like *Why Not South Nagaland*, *Nagalim for Christ* and *Israel in India*. These books were aimed at propagating the idea of a separate homeland for Naga tribes residing in Manipur, Assam and neighbouring Myanmar. These books were banned by the Governor soon after their circulation. But many had gone through these books.

It was inevitable that other ethnic groups like Kukis and Meiteis would have fiercely opposed such an idea. Hence, Naga–Kuki clashes during my second tenure in Manipur did not come as a surprise to me. What was surprising is that the authorities took no action at all even after such ominous indicators were available for long. To add oil to the fire, it was also widely believed that some elements in the government machinery had made efforts to arm, train and support the Kuki militants against the NSCN. Other minority groups like the Nepalese living in the tribal areas were coerced into supporting their respective 'masters', Kukis or Nagas in whose land they were tenants or settled. Many of the Nepali youths were deployed to protect the Kuki villages too in return for their continued privileges in what they called their land. The so-called Nagalim (Naga Country) guards fully armed by the NSCN were tasked to protect the Naga villages. Playing directly or indirectly one ethnic group against the other in the garb of gaining equilibrium as propounded by some in the security establishment even to this day is a sure recipe for disaster.

I still wonder if this was a simple ethnic conflict. Or was it ethnic cleansing? A civil war? In one instance, one Kuki chief of Khonglo Vaiphei sold the entire village land to a neighbouring village chief and left the village to secure the safety of his family. Most other villagers were left landless. The villagers had to be accommodated in relief camps. Even the relief camps established by the authorities were shoddy. As per Kuki customs in Manipur, land of the village belongs to the chief not to the individuals. The same custom does not apply for Kukis in other states like Nagaland or Mizoram. The scars on both sides as well as on the idea of Manipur will last longer than the survivors of the conflict of 1992-93. President's Rule was imposed on 31 December 1993 and an elected government headed by R.K. Dorendro Singh was dismissed. The continuance of Naga-Kuki ethnic strife shall be dwelt upon again later in the light of the Framework Agreement between Government of India and the NSCN (IM). I am happy that I could contribute significantly to the efforts of the military and the Government in containing the ethnic conflict and in bringing about sanity to the people engaged in the carnage.

Meitei and Meitei-Pangal (Manipuri Muslims) clashes during

1992 was another very unfortunate and avoidable incident which occurred during my tenure. It all started over rumour of a Meitei family including children having been burnt alive in their car by some Pangals. The Pangals are believed to have been brought to Manipur by the kings after the campaigns in Sylhet (in Bangladesh), Tripura and Cachar region to employ them in various military related works in the Manipuri army around the 17th century. The king allowed these Muslims to be naturalised and were given *Yumnaks* (Manipuri clan titles). Many of them married local women. The arrival of Shah Shuja, brother of Mughal Emperor Aurangzeb, along with remnants of his retinue in Manipur to take shelter under the Manipuri King during 1661 also could have resulted in some Mughal influence among the Muslims here. Shah Shuja was in self-exile in Manipur after the King of Tripura supposedly advised him to shift his location from the Arakans to Manipur to escape the onslaught of the Mughal army and spies on the lookout for his party ('Pangal', Wikipedia). Meitei Pangals, therefore, have adopted a mix of Mughal, Bengali Muslim and Meitei cultures. The Pangals follow many Meitei social practices like wearing of 'Phanek' (sarong) by the women, *Maiba–Maibi* system (traditional healing) and so on. Sandwiched between their origin and the Muslim faith, I believe that some in the community are facing identity issues at present. In trying to move away from being Meitei-Pangals, many of them have discarded their belief in the naturalised status. Religious practices which originated from the Arabs have been adopted by many of them thus alienating them further from the culture of Meitei society. The community comprising over 9 per cent of the population in the state has served Manipur very well in every sphere. They have excelled in many fields. Today, many in the community feel the need for asserting their perceived rights as indigenous people. In fact, the Meitei Pangals are now demanding ST (Scheduled Tribe) status for Muslims of the state.

In the instant case of violence between the Meitei Pangals and some Meiteis, the rumour travelled very fast and some miscreants started swift retaliation from both sides. Within no time, my Brigade Commander Brig. Patankar and I drove through the affected areas. The Commander ordered deployment of troops in aid of civil authorities alongside the civil police. The clashes lasted about 72 hours including isolated incidences in the remote areas. During the night, I was deputed to liaise with the police regarding the operation to prevent further escalation of the communal clashes. Another dark chapter in the modern history of Manipur indeed.

As per population census of 1931, the population of Pangals was 22,864 only out of total population of the state of 4,45,606 (approximately 5 per cent). It is estimated that the combined population of Pangals and other Muslims who have come and settled in Manipur from Bangladesh or Assam number 2.4 lakh, as per census of 2011 (8 per cent). In the aftermath of this unfortunate event, a few Meitei-Pangals formed terrorist groups with the assistance of ISI agents. The groups which established their presence with the help of ISI during 2002-10 are Islamic Liberation Army (ILA), People's United Liberation Front (PULF), and Muslim Volunteer Force (MVF). Most of them had linkages with Muslim fundamentalist organisations in Assam. There are no reports of violent activities by such groups today. But, the seeds of conflict remain. Reports suggest that radical elements are active in Manipur as it is in other parts of the North East Region. The possibility of lingering communal animosity cannot be wished away. There are also unconfirmed reports about huge number of Muslim immigrants from Cachar in Assam and from Bangladesh in the recent past, particularly in the Valley districts and Jiribam district. Many Rohingyas from Myanmar are believed to have found their way into the state via Bangladesh route during the last few years. It is a challenge for the administration as well for the community to analyse the effects of such illegal immigrations into the state. The demography of the state is in a fragile state. Any disturbance in the existing profile may have serious consequences in the society. The Pangals of Manipur will find it difficult to shed their Pan-Islamic identity whereas their coexistence with the Meiteis could be positively influenced by their acceptance of their local version of indigenous Muslims known as 'Meitei Pangals'. My role in this dark episode was limited to work with the state machinery and the military to quickly contain the violence. Suitable mechanism involving Civil Society Organisations and inter-faith understanding is imperative to prevent re-occurrence of conflicts of such nature in future.

Deadly Ambush by the NSCN

As stated earlier, NSCN had carried out major ambushes on the Security Forces (SF) in Chandel district during 1992-93. Besides putting pressure on the SF, these were intended to warn the Kuki groups as well. On 29 June 1993 a Company of 15th Battalion of the Assam Regiment was moving on Sita-Tengnoupal road in southeast Manipur. At about 10 a.m. a series of loud explosions followed by small arms firing was heard by me on that day while I was driving from Pallel to Tengnoupal on a mission. My driver and I drove towards the site of ambush and reached there in about 20 minutes from the highway. Isolated firing was going on when we reached the spot. But the Company had suffered 16 dead by the time we reached and a few others injured. Troops who survived the attack were visibly stunned. The Company Commander who was also the convoy commander had just crawled up the slope and came to the road where I met him. He regained command and control soon and ordered his men to search the site. But the NSCN ambush party had melted away in the jungles soon after the ambush leaving two of their cadre also dead.

As an infantry officer I observed that the ambush was meticulously planned, tactically sited, and swiftly executed. A text book style ambush. We failed to read the ground and the situation. The killings were savage. I saw one of the dead bodies highly mutilated inside the engine compartment of the Shaktiman truck. Once again the follow-up action by the army was intense. The SF agencies gradually but surely traced the planning and orders for the ambush, allegedly linked to some politicians. But the evidence was not adequate to frame these culprits. The army couldn't achieve much to nab the culprits. It was a dirty war. Overall, the activities of the Naga, Kuki, and Valley-Based Insurgent Groups were at their peak during the period. The activities of Naga and Kuki groups were aimed at consolidating their identity and their idea of homeland. Reports of support to the groups by elements in China and Bangladesh were worrisome indeed.

The Valley Armed Groups

In the meanwhile, the valley armed groups were in the process of regrouping and forging re-alignments between the various groups and factions during the period. UNLF (United National Liberation Front) cadres were being trained, and the first batch comprising 34 trainees 'passed out' on 9 February 1987 under Chairman Sanayaima in NSCN base area. This day is also 'celebrated' as the Manipur People's Army (MPA) Day. Efforts to unify the various groups were made during the period but bore little fruit. Factions within the NSCN during the period also played a role in the process. Formation of the Indo-Burma Revolutionary Front (IBRF) on 21 May 1990, with NSCN leader S.S. Khaplang, ULFA chief Arabinda Rajkhowa and Sanayaima as 'Chairman', 'Vice Chairman' and 'General Secretary' respectively did not make much headway due to various factors. However, UNLF had gained significant ground in southern part of Manipur by 1994-95. The Indian Army was successful in driving away the UNLF bases after a hard campaign from 2003 to 2005. PLA (People's Liberation Army of Manipur) played a role in containing the communal violence which erupted between the Meiteis and the Pangals.

I found that our world in Manipur and in the region was in constant struggle between modernity and antiquity, between the haves and have-nots, between newly evolving democratic values and the values of the past. Not to mention between Pride and Prejudice! The way forward appeared to me was to stress on more and more education, cater to spiritual and emotional needs of the people and of course the governance of the day.

In the Land of Babri Masjid

My second tenure in Manipur came to an end in July 1993. I joined my unit now located at Faizabad, 10 km away from Ayodhya. I reached the unit location soon after the December 1992 demolition of the Babri Masjid there. I was very keen to have a look at this famous site. Outwardly, it appeared to me that the situation was under the firm control of the administration. It was an unbelievable sight to see remnants of the historic masjid in ruins. It occurred to me many a time that the history of India has evolved after many such acts of violence, wars, and struggles among its peoples for centuries. Many people in the area believed that the demolition of the structure was an act of God. Faizabad was also known for a Mauni Baba who many years back stayed in a hut on the banks of Saryu River. The Baba was suspected to be none other than Netaji Subhas Chandra Bose in hiding after India won Independence in 1947. I visited the site of the hut too.

Col (later Gen.) V.K. Singh was still the Commanding Officer of the unit. It was a great learning experience to serve with him again. The unit routine involved training activities for war. Mangala and daughters Poornima and Priyanka joined me at Faizabad during mid-1994. She left her job in the Assam Rifles temporarily. In his book *Courage and Conviction* Gen. V.K. writes about 'Fear and Panic'—an episode attributed to me. The unit was conducting a Promotion Cadre to promote Havildars to the rank of Naib Subedars in the unit. The CO, Col V.K. visited one of the classes in progress. I accompanied the CO since I was the Officer in Charge of the cadre. The topic of discussion in the class was 'Fear and Panic'. Accordingly, the CO wanted to test the level of understanding of the candidates. He asked for any volunteer to explain the difference between 'fear' and 'panic' in battle situations. Prompt came the reply from an enthusiastic candidate. He replied: '*Sahib, Fear hota hai Dar. Jo niche hum logon-ke level main hota hai. Aur panic hota hain unche darje ka dar. Jo officer level main hota hai*' (Sir, fear is Dar which can happen to lower ranks, while panic means a higher state of fear which affects officers at higher levels!) It was an awkward moment for me because I had myself explained the subject in Hindi an hour back. I had explained to the candidates in Hindi that 'panic' is only a higher state of fear. The candidate perhaps took it that 'panic' is always there at higher levels! And 'fear' was what happens to the lower ranks!

Corridors of Power: Lutyens Delhi

In the month of June 1995, I was to report to the Army HQ as Assistant Military Secretary in the Military Secretary (MS) Branch dealing with Postings and Transfers of officers in the rank of Colonel in Staff and Instructional appointments. It was a truly prestigious post for a Lieutenant Colonel. I still wonder how I could have been assigned this onerous responsibility. Later, I learnt that my unit colleague Col (later Lt Gen.) V.K. Pillai, my coursemates P.M. Hariz (retired as Army Commander), Narender (retired as Brigadier in the Corps of Engineers) who were in the same branch did put in good words for me to the authorities. I was told that MS Branch preferred officers with highest integrity and unblemished spoken reputations. The assignment was a very challenging one. My primary role was to manage transfers and postings of all officers of the rank of Colonels on Staff and Instructional assignments which included the cream of the army Colonels who exited from the prestigious Higher Command (HC) and Long Defence Management Courses (LDMC).

It was in this tenure I learnt that one could make a difference in the environment even as an officer in the junior ranks. One must be logical, fearless in putting across one's views, accept other's views when called for with a force of a better logic. As a junior functionary in the MS Branch, I was not authorised to interact with the officers of the rank of Colonels whose postings I was to deal with. I was required to process posting proposals keeping the records of service in mind. It was easier said than done. Records don't speak much of the people concerned. Records reflect the officially documented papers of the concerned officer. Ninety-nine per cent of the Army officers have clean records without any adverse disciplinary or administrative records. Hence, it was a challenge to identify officers for a particular post purely based on records maintained in the HQ. In other words 'Right Man For The Right Job'. Therefore, I reasoned with my bosses that it was necessary to interact with the officer before his transfer orders were issued. My bosses felt that this would only raise hopes and aspirations among the officer cadre which we would not be able to meet. The Additional Military Secretary, Maj. Gen. D.S. Chauhan saw my point and he encouraged everyone to partially open up the secrecy surrounding the posting processes in the MS Branch. I am glad that today very interactive and open systems exist as far as postings upto the rank of Brigadier is concerned.

The working environment was primitive. Our office in the basement of the South Block was once the stables for the army horses during the British period. We often had to chase away monkeys from our offices, particularly on Sundays when the monkeys entered our office to rest. We used to buy quality file covers from our own pockets if a

File Noting had to go right upto the Military Secretary or the COAS. In sum, I felt that Army HQ needed a revamp in its structure and staffing pattern.

In a lighter vein, on one occasion, I politely asked my boss if he had a 'Buddy' (A Military help authorised at certain levels for uniforms and personal assistance) at home pointing to the sorry state of the uncleaned brass on his shoulder. He had obviously not liked a junior officer checking his dress and turnout. A thorough gentleman that he was, he conveyed to me that I along with my family were invited for a dosa session at his house on the following Sunday since I had checked his turnout! I gladly accepted the offer of some beer and dosa! We enjoyed the pleasant afternoon at their place. My boss's charming wife then brought the same brass rank epaulettes and buttons to show me and said that these could not shine any further than they already were. I was totally astounded when she told me that those were made of gold. Therefore, it could not shine any further! A real lesson for me to avoid commenting on anything unless details are known to me!

3

World's Highest and Coldest Battlefield

A Dream to Command a Fighting Unit in War

I was cleared for command of a unit in the rank of a Colonel. I was told by Col Atmanathan, the Colonel MS (Infantry) who was responsible for postings of Infantry Commanding Officers, that it was not possible for me to wait to command 2 Rajput, the unit in which I was commissioned. My seniors Pillai and Amulya Mohan were also keen to command the same unit. Hence, I assumed the command of the youngest unit of the Regiment, 27 Rajput on 5 September 1997, popularly known as *Sattaees* a new raising but 'battle hardened' by the experiences in the Kashmir valley during 1992-93. The unit was due to move to Siachen Glacier in a few months from Delhi. It was a lifetime opportunity and a great privilege to be in command of such a fine set of soldiers in the regiment. I was lucky.

The tenure in Delhi for the unit was a mixed bag. Troops belonging to

NCR (National Capital Region) and nearby places were able to attend to their family commitments. Facilities to look after the children's education and health of the family members were available. The unit was able to contribute to various events of national importance. But the downside included issues of manpower commitments of all hue which affected unit training. Maintenance of military good order among troops in the environment was also a challenge. To balance the welfare of troops and the official and some semi-official responsibilities was the key. My unit being a new raising did not have a comfortable bank balance in the Regimental Fund Account. My endeavour was to recover whatever debts other agencies and officers owed to the unit before the unit moved to the next location. On one such an attempt, Major J.L. Todkar, of the unit informed me that the GOC of the formation who was my Reporting Officer owed the unit some outstanding bills. This was due to a private get-together which the GOC hosted in our mess at Delhi. He was otherwise based in Meerut. The 2IC (Second in Command) suggested that we write to all officers who have not cleared their mess bills whether they were from our unit or otherwise. I took the bait and wrote a Demi Official letter (a letter that communicates official mail along with personal news) to all including my own Reporting Officer (GOC)! It was a big mistake on my part to have written a DO letter to the GOC for this purpose. Later, I realised that the right approach should have been to walk upto the GOC personally and request him to help the unit to build up the funds by clearing his bills. He would have understood and this embarrassment could have been avoided. The bill due from the GOC was cleared immediately but with a terse comment that perhaps I must have been the only CO in Army in recent times to have written a DO letter on such a subject to his own General. He obviously had not liked it. On introspection, I felt that I could have conveyed to him the same thing differently in a more dignified manner not to hurt his self respect and ego, in person in private. I am sure that he would have appreciated me had I conveyed the same in a one-on-one meeting. As expected the Annual Confidential Report initiated that year on me almost sealed my career. I knew that I had rubbed the superiors the wrong way by some of my actions and omissions such as above. My Brigade Commander at that time had also made known his mind on similar issues. With my experience, I sensed that it was my last rank in the army!

My unit set course from Delhi to play our part in the Siachen Glacier in early 1998. Medical category officers were transferred since they are allowed to serve in such areas only on a voluntary basis. One of my most memorable rail journeys was the unit special train to Jammu from Delhi along with some of officer's family members including mine. We took three days to reach Jammu in the special train but it was well worth it since the ladies and children enjoyed it. We took darshan of Vaishno Mata at Katra and dispatched our families back to Delhi. Major Kuldeep Mehta, one of the finest officers of my unit, had arranged for a good reception. We proceeded with the journey to our destination by road. Ladakh in the summer months is literally a heaven in the high altitude region. This being the first high altitude tenure of most of the troops, we had to be very careful in planning any activities for them.

I decided to cross the Ladakh Range in a cross-country trek with maximum of troops of my unit on foot from Leh to Nubra valley as part of our induction to the glaciers. We commenced our trek on 3 May 1999 flagged off by the GOC from Leh. My Adjutant Capt. Negi was with me in the trip. Such was the ignorance and intelligence set up at all levels about the impending war (Kargil) that my battalion which was the Quick Reaction Force of the formation was on a trekking expedition across the Ladakh Range just three days prior to the official declaration of the war! Unthinkable but true. 'Operation Vijay' officially commenced on 6 May 1999. Heavy unit stores were dispatched by road through the famous Khardung La Pass. We undertook a three-day trek across Digar La Pass in Ladakh Range to descend to the Nubra valley. I learnt from the civilians in the area that this Pass was made use of by the Indian forces during the 1948 campaign against the Pakistani army. We noticed many bones of animals and may be humans too en route to the Pass. The road through the Khardung La Pass came much later. The top of the Ladakh Range afforded the most gorgeous of panoramic views of the Zanskar Range across the Indus to the south and the enchanting Nubra valley to the north. Trekking in the Himalayas, in its valleys, passes and mountain tops, has been truly a joy for me. I only wish that the future generations too find the same glory and beauty in the hills and mountains of the Himalayas a century later.

On reaching Nubra, 20 of my men along with me rafted down the Shyok River towards Chalunka, quite oblivious of the developments in Kargil. Perhaps it was the first ever rafting in the Shyok River. We did not follow up to claim this achievement as the first of its kind. Many years later in 2008, one artillery unit did a similar adventure and claimed it as the first. The Brigade Commander came to receive us. He also conveyed that things were bad in the neighbouring brigade in Kargil due to some infiltrations having taken place. He also advised me to be ready for any task that may come to my unit.

The Kargil Conflict as I Saw It

'Where Courage and Fortitude is the Norm'

'Operation Meghdoot' was launched on 13 April 1984 under the command of Brig. Vijay Channa, Commander 26 Sector. The backdrop of 'Operation Meghdoot' is well known to the readers. In short, it was a race between the armies of India and Pakistan to occupy the Saltoro

Ridge which dominated the 75-km-long Siachen Glacier. Along with other glaciers in the area, it occupies nearly 3,000 sq.km. area. NJ9842 was the last point upto which the ceasefire line was described and annotated in the aftermath of the Karachi Agreement of 1949 between the two countries after the 1948-49 conflict. The Karachi Agreement described the ceasefire line upto NJ9842, beyond which the line was to be 'running northwards to the glaciers'. India describes the northward line to be along the watershed of Saltoro, whereas Pakistan described the line as one joining the Karakoram Pass and NJ9842. It was only in 1985 that an Indian survey team identified map point NJ9842 with an approach from Urdolep glacier. My friend Capt. (later retired as Lt Gen.) Sanjay Kulkarni and his team from 4 Kumaon were the first to land at Bila-Fondla on 13 April 1984. Thus began the Siachen saga, the highest battlefield in the world.

The Pakistani army was planning a major push towards the Turtuk Sector close to NJ9842 during the Kargil War 1999. This sector is more than 100 km away from the Drass area where the battles of Tololing and Tiger Hills were fought and seen in each and every household TVs of India. Unknown to our commanders and staff at every level during the period 1998-99, a determined Gen. Musharraf was also infiltrating troops across the LoC in this sector. During the autumn of 1998, we had received a number of intelligence inputs in bits and pieces but most of them were found confusing and contradictory at the unit level. The inputs pertained to buildup of Pakistani forces across the AGPL (Actual Ground Position Line which generally demarcated the boundary beyond NJ9842) and LoC (Line of Control which delineates the boundary from the International Boundary in Jammu region upto NJ9842) in the vicinity of our division. A clear picture of the situation was not available at the ground level. Overall, we assumed that something serious was in the offing. But none of us in the unit or higher HQs could assess the real intention of the adversary. The visit of the Pakistan Army Chief to a brigade HQ located at Dansam just across the LoC twice in a matter of three months itself was a big battle indicator that something important was being planned sooner than later.

In retrospect, I am of the opinion that the entire intelligence machinery in the country had collapsed or their priority lay elsewhere. The army command hierarchy did not fare any better. Lt Gen. S. Padmanabhan, the Northern Army Commander, was transferred two months prior to the operation. I never saw nor met the Corps Commander Lt Gen. Krishan Pal in my 24 months in the Chalunka/Turtuk/Siachen area nor during the Kargil conflict. The Corps Commander was the Senior Reporting Officer for all the Commanding Officers of major units like an infantry battalion in the chain of writing the confidential reports which decides the career prospects. It clearly showed where the priorities were. Higher headquarters only forwarded the hundreds of intelligence inputs in bits and pieces without any coherent analysis. I often wondered what the layers of intelligence units and the staff up the chain were doing. At the unit level, we were mostly unaware of the happenings in the higher headquarters. In any case I suspect if anything worthwhile was even known in the chain.

At least on one occasion I met the Colonel General Staff at the Division Headquarters responsible for operations during the run-up to OP Vijay to find out the latest situation. He was no wiser than I was on the latest input even a few weeks prior to the unfolding of events of OP Vijay (Kargil). I recall that we had a discussion on operational matters of the Division during March/April 1999 in which all the formation commanders and the Commanding Officers of the formation took part. The possibility of the scale and the kind of Pakistani intrusion that actually took place was never seriously considered a possible threat scenario. I, as CO, was not privy to what was being planned at the higher headquarters to counter the threat. But my focus on training of small detachments in the unit resumed with much greater focus and urgency. Operating in small teams is the only way to fight in the super high altitudes that we were operating in. The pattern of enemy artillery shelling, helicopter flights, logistics buildup, routine in the posts of the adversary during end 1998 made me stronger in the resolve to keep my unit fully ready for any eventuality. Soon we were relocated to Chalunka valley in the first week of August 1998 where exchanges of artillery fire between the Pakistani army and our troops were routine. My officers and men were fully prepared by the time action started in some areas in Kargil sector in May 1999. There was increased enemy artillery shelling in and around the Chalunka Sector (This sector was recaptured from the enemy during the 1971 War).

On the night of Handing/Taking over with the Southern Glacier unit of that time, Chalunka was heavily bombed around the first week of May 99. I was with my Second in Command Lt Col Bhattacharya, when one of the Repair and Maintenance vehicles had a direct hit by one of the enemy shells. We were about 100 metres away. Two of our EME (Corps of Electronics and Mechanical Engineers) soldiers died on the spot. I ordered my unit which was located close to Chalunka and under tentage to relocate to another location 5 km away to be outside the artillery range of the enemy. In an hour the Pakistan radio announced in Balti language that a Rajput unit has run away from the area! It was strange that the relocation of the unit was known to the enemy almost in real time. It was obvious that elements of the enemy were already in the vicinity. In the meantime, a large quantity of arms and ammunition were intercepted in Turtuk Village by our forces. A few of the villagers of Turtuk had obviously been subverted. Turtuk is the last village in Shyok Valley. A scenic village known for apricots and walnuts, it is inhabited by a mixed ancestry of Tibetan and Indo-Aryan descent speaking the Balti dialect. Today, the village is one of the tourist hotspots in Shyok Valley. The entire stretch of about 60 km of LoC between the Turtuk Heights and Chorbat La on the Ladakh Range was rugged and very difficult terrain. The area was regularly patrolled by our own troops. In the winter of 1998, we did carry out more intensive patrolling in these heights. In his book on Kargil, General Musharraf had made a mention about aggressive patrolling by Indian forces during the period. 'Operation Vijay' had commenced on 6 May 1999.

It was on 8 May 1999, that my superior Brig. (later Lt Gen.) Katoch conveyed to me that something very serious was happening in Kargil and that my unit should be prepared to proceed back to Kargil to restore the situation if the need arose. Things happened very quickly. It became clear within a day or two that the Pakistani army had intruded into our territory in Drass and Mushkoh valley in Kargil area. The quantum of intruders and their intent could not be assessed at that point in time. Operations to evict the intruders by our troops had begun near Drass. I was surprised by news reports that an IAF fighter plane was shot down by the Pakistani forces on 27 May 1999 in the Kargil sector. This piece of news was a very important indicator that we were likely to go to war soon. All attention of the hierarchy went to Kargil area to evict the intruders. I was later told that my unit was to continue to operate in the Chalunka sector across the Khardung La Pass which connected the Nubra valley and Siachen area with Leh, and to relieve 5 Para from the glacier who were being re-deployed at Kargil along with elements of 8 Mountain Division. Accordingly, we were in the process of relieving the Paratroopers from the Chalunka and South Glacier area during the second week of May 1999. I have very fond memories of the de-inducting unit since I spent many weeks in the glacier with them while they were deployed in the glacier. I was personally present to wish them luck when the unit was being heli-lifted to Kargil. One of the soldiers presented me with a rosewood cane on his way to the helipad. He also confided in me that enemy troops have infiltrated the area and that my unit should be extra careful there!

Earlier, 56 Brigade had been inducted to Drass from Srinagar valley. Entire 8 Mountain Division from the valley had been inducted north of Zojila Pass to evict the enemy from Mushkoh, Drass and Kaksar sectors. The 3 Infantry Division, to which my unit was orbatted, was tasked to restore the situation in Batalik and Turtuk sectors to the north and northeast. Tiger and Tololing Hills overlook the village of Drass and the Srinagar-Leh Highway. Isolated actions at various points had begun along the LoC from Mushkoh to Chorbat La, a front of approximately 50 km. Media was in full swing to broadcast the Bofors in action, troops buildup, and troops approaching the Tiger and Tololing Hills where the Pakistani troops had already occupied and deployed themselves seven to ten km deep inside our territory. The media presence in the proximity of the battle zone was unprecedented. The unifying response of the nation during the Kargil war owed much to the media. For the media who were mostly deployed near the Drass highway, the war was just another story. At the end of the war the Indian forces were able to evict the intruders from the area. The Indian Army had 474 soldiers killed and 1,109 wounded. I consider that it is not for me here to discuss more than necessary on the conduct of the Kargil war as a whole and the operations in Drass, Mushkoh, Kaksar, Batalik and Chorbat La sectors of the war. I have only tried to put down what my proud unit 27 Rajput did during the operations in the Turtuk-Siachen sector.

Our Attack on Point 5770 (Enemy Balal Post) near NJ9842. OP Vijay (Kargil)

By mid June 1999, the situation appeared very grim for us. Unknown to the media or to some of the higher authorities, a similar situation was unfolding in areas close to the Siachen Glacier north of Chorbat

La in the Ladakh Range. One such feature of significance was Point 5770 near NJ9842 from where the boundary was never demarcated on ground after the Simla Agreement of 1972. My unit was initially given the task through a telegraphic message to 'occupy unoccupied mountain peaks' in the area. However, we were to focus on occupation of Point 5770 feature as early as possible. This feature is located in proximity to NJ9842 across the Shyok River from where the AGPL (Actual Ground Position Line) alignment begins. My own physical reconnaissance of the feature from very close proximity (along with Major Saini and Capt. Shyamal Sinha) during the period 10-19 June, and other inputs indicated that Pakistani troops were already in occupation of the 19,000 feet high mountaintop. We were able to get close to the objective at night and it enabled me to make a detailed assessment and plan to destroy the enemy. I conveyed to the Brigade Commander that it was going to be an attack operation before going in to 'occupy the feature' as directed and not just to go and occupy the feature. He was visibly unhappy about the unexpected piece of information about the presence of Pakistani troops on Point 5770 feature. No further orders came in writing but he kept insisting that we should 'occupy the feature earliest'. The higher headquarters helped us with a four-member Rope Fixing Party from the High Altitude Warfare School led by Capt. Shyamal Sinha. Thereafter, I instructed my Adjutant Maj. Negi to inform in writing to the brigade headquarters that we were going in for a 'silent day attack' during the end of June 1999. The risk involved in my attack plan was huge. But the element of gaining 'surprise' was paramount in my mind. Loss of surprise would have jeopardised the entire effort and my troops could have been sitting ducks in the hands of the enemy. I took the calculated risk anyway. And the Brigade Commander agreed to the plan. It appeared to me that most were quite confused in the beginning on how to deal with the situation at hand. Body bags at Leh Airfield told the story of disorganised response by our forces. Besides the airstrikes and artillery guns booming in the media, not much was achieved on ground till the fall of Tiger Hill on 23 June to our forces.

Preparations for the attack on Point 5770 were in full swing from 11 June onwards. All my company commanders, Maj. (later Brig.) Vikram Yadav, Maj. (later Brig.) S.K. Saini, and Maj. (later Brig.) S.K. Saul volunteered to lead the attack. I selected Maj. (now Col) N.S. Cheema as the Commander of the assaulting troops since the entire operation had to be carried out at Company level and he was the only one to have received formal training in high altitude warfare course. Capt. Shyamal Sinha was the Commander of the Rope Fixing Party. Maj. Saini who was the Company Commander in the area was made in charge of all support troops including artillery fire control. Ropes had to be fixed stealthily till a place beyond the halfway climb to the objective. Rope-fixing is a mountaineering technique which involves a very slow and deliberate action to fix the rope on the extremely steep rocky or icy surfaces by using specialized mountaineering equipment like rock or ice pitons. The rope-fixing was done under the cover of darkness or in extreme bad weather conditions to conceal our activity. The pitons had to be padded with thick cloth to reduce the noise caused by hammering the pitons into the rock or ice surfaces. These ropes would then be used by the assault troops who have to carry heavy weapons and warlike stores for the attack on the enemy. Two observation posts were established to keep surveillance on the enemy from the flanks while the rope fixing was in progress over a period of three nights. The Reorganisation Group along with the requisite stores was also ready at the base of the objective in the glacier itself. There was no dearth of volunteers for the attack among my troops. It became difficult for me to send back troops who were keen to be part of the assault, due to their limited exposure in this super high altitude.

I had chosen my command post at a distance of about 700m from the objective from where I could observe our troops as well as the enemy troops deployed at the same time. The Command Post was nothing but a position below a rock in the glacier with communication facility. Our troops would not have been able to see the objective till they were only a few metres away from the enemy sentry post. It was also difficult for the enemy to see the assaulting troops since the gradient and the route chosen was extremely difficult—at 80° incline with hanging cornices, ice gullies and big rocks along the route of the climb. Therefore, my party comprising of self, my runner Sepoy Madan Singh (later retired as Honorary Captain), and a radio operator, was the only group to observe both the enemy troops and our assaulting troops simultaneously. We decided for a daylight assault due to the nature of terrain. It would have been almost impossible to climb the feature at night. We also had the advantage of bad weather conditions. A silent assault was planned to maintain surprise during the operations. The first attempt to capture was made on 25 June 1999. Accordingly, operations began at 0100 hrs on 25 June 1999. However, we had to abort the mission due to extremely hostile blizzard conditions which enveloped the area around 11 a.m. Blizzard conditions (severe snowstorm with high velocity winds) at such altitudes are extremely dangerous and life threatening. The second attempt was planned for 27 June on the advice of one of my trusted JCOs, Subedar Brijlal, who advised me that 27 June being a Tuesday, Lord Hanuman's day and combined with the number 27 of our unit, this day was most auspicious and that the attack will be a grand success. He volunteered to be part of the assault but I decided against sending him due to his advanced age. There was hope and prayers among each one of us for the success of the attack mission.

We launched the second assault on 27 June at 0215 hrs with Major Cheema in command of the Company, minus a platoon force excluding the support elements. The day was also the birthday of Prophet

Mohammad. The weather was calm. The fighting in the Drass sector was in full media glare since parts of the operations were accessible to the media who were strung along the Srinagar-Leh Highway. The events unfolding across Khardung La were out of reach for the media. The greater attention of the army hierarchy was also obviously with where the media was. Small-scale actions in the Chalunka sector (across Khardung La) had commenced during the same period i.e. end April/May 1999. In one of such encounters, Capt. Hanif Uddin of 12 Rajputana Rifles was killed. Pakistani forces were active near the southern glacier, Turtuk and north of Chorbat La. The attack on Point 5770 thus went unnoticed as well.

The daylight attack went with clockwork precision as planned. I was able to observe each one of the assaulting troops from my position. Major Cheema and Capt. Shyamal Sinha were in constant communication with me through the telephone line which was being laid as the attack progressed. Telephone line connectivity is a preferred means of communication during such sensitive times. The enemy troops could be seen doing their routine activities. I saw a sentry who was observing towards the direction of the attacking troops. But the gradient was so steep that the enemy troops could not observe our men. At times, I had to telephonically guide the troops on the handset, step by step to negotiate the boulders and hanging ice cornices above them. Capt. Shyamal Sinha, whose party was hardly 50 metres below the objective, told me that he was unable to negotiate a big boulder ahead. Luckily, I was able to guide them to negotiate the same to maintain the momentum of attack! It was at about 1.30 p.m. that both the assaulting parties led by the two officers were just a few metres below a lone enemy sentry, a little away from his weapon on a rock. The lone enemy sentry was seen either writing or reading something. I told Cheema and Shyamal that I would communicate with them only after the fighting was over. I wished them all the very best and to be safe. The situation could not have been any better.

God was kind. I continued to observe and monitor them. The first person to approach the enemy sentry was Capt. Shyamal. On seeing the sudden appearance of the officer, the enemy sentry rushed towards their fibre glass hut (This was not visible to me being on the reverse slope, but narrated later to me by Shyamal and Cheema). Thereafter, I heard loud noises of firing and grenades. I prayed for the safety of my troops. Cheema's LMG (Light Machine Gun) did not work and could not fire at the enemy. The PIKA gun of the Pakistanis in the abandoned sentry post was used by Naik Pahal Singh (he later retired as Honorary Captain) for fire support during the fighting (This gun is a variant of German-made MG1A3 Pakistan ordnance factory made, called PIKA gun by us). A few Pakistani troops were seen abandoning the post and running away towards their side. The faint echo of the battle cry of the Rajput regiment *Jai Bajrang Bali Ki Jai* could be heard at intervals. I knew that our boys have done it for sure. I prayed for the well-being of my boys whom I had sent in harm's way. By 2.30 p.m., Cheema reported to me the 'SUCCESS' signal on line. And most importantly all our soldiers were safe. Ten enemy soldiers were killed. Within minutes the Pakistani artillery opened up with the ferocity I had never seen before. Hell broke loose thereafter. We counted more than 400 rounds of field and medium guns falling in the vicinity where my party was. One round fell a hundred metres from me. An earth-shaking and near-death experience indeed! Major Saini ensured that our own heavy guns including Bofors retaliated and silenced the enemy guns. Saini also ensured safety of own troops elsewhere in his area. The artillery duel continued for most of the day and night. Airburst shells (a shell or grenade that detonates in midair) were fired on our location through the next few days and nights. One of the enemy air burst shells injured my radio operator who was just a few metres away while resting under a rock. One M-11 marked missile of 2.5 metres length fell a few metres from my unoccupied bunker. We were not able to identify the type of missile till this day. The enemy's efforts for a half-hearted counter-attack failed with more casualties on their side. We suffered two casualties during this period of consolidation, after the physical capture of the objective.

Nothing Succeeds Like Success: The Other Side of War

The dead enemy soldiers were buried under the snow with due military decorum on the objective (Point 5770) itself after the Pakistanis refused to accept the bodies as their own. In his book *Beyond NJ 9842: The Siachen Saga,* Nitin Gokhale, defence analyst, writes:

> None of the media people could manage to get to the conflict beyond Partapur. Most of us concentrated at the easily accessible Kargil-Drass sector and reported from there. The Chorbat La-Turtuk sectors were totally under reported. But away from the media glare, and dare I say even from the gaze of the higher military authorities too, a major victory was achieved by 27 RAJPUT.[1]

A month later I was asked if we had the body of one Captain Taimur Malik of 3 Northern Light Infantry (3NLI) of Pakistani army. I replied in the affirmative. My unit was told to exhume the body of the officer to be evacuated to Drass area. I strongly opposed the idea of sending only the body of the officer and suggested that all the dead bodies be exhumed and handed over to the Pakistani army. The higher headquarters relented! I took the risk of danger to my troops who would be involved in the recovery of the bodies. We required 7-8 personnel to evacuate one dead body from such an altitude. My troops

1. Nitin A. Gokhale, *Beyond NJ 9842: The Siachen Saga* (Delhi: Bloomsbury, 2014), p.203.

did it. Four bodies were airlifted along with the body of Taimur Malik to Drass and later handed over. I went through the letter written by the father of the slain officer found in his shirt wherein he motivated his son to fight and becomes a 'Ghazi'. The father, a retired Brigadier in the Pakistani army, wrote to his son, '*I fought the Indians in the 1971 War. You should fight them and become a Ghazi yourself.*' Another letter written by the mother of the same officer asked her son to get back home soon and that she would arrange his posting to a better place! Capt. Taimur Malik's letter to his father which was in his pocket was later sent to his father as a goodwill gesture.

It was a great success which will remain in the annals of military history of India. A daylight assault, a silent assault where not a single artillery (heavy military ranged weapons) shot was fired just before or during the attack. This was one attack in the entire Kargil war where no artillery guns were used and where no casualties to own troops occurred inspite of the most unfavourable terrain and challenges of nature besides the vantage position to the enemy. Failure of the attack or loss of surprise in the operation would have been a disaster for my unit, and I would have had to bear the full consequences for the same. The entire risk was mine and mine alone. Leaders are required to take risks when the time comes. It was an action which prevented any Pakistani designs in the Chalunka Sector which could have threatened our position in the glaciers. In the words of Maj. Gen. Ashok Kalyan Verma who authored the book *Kargil: Blood on the Snow*, the capture of Point 5770 by 27 Rajput under my command was 'perhaps the most incredible action of the Kargil war' (*Kargil: Blood on The Snow,* Delhi: Manohar Publishers, 2002, p. 125). My unit also occupied some other unoccupied heights and mountain peaks above 18,000 feet in the north of Shyok River.

One lost opportunity one year earlier was our attempt to occupy Thang Top in the mouth of Shyok River as it flowed to POK. The LoC in this area as defined in the Simla Agreement 1972 includes Thang Top as one of the coordinates on the LoC. My troops were successful in reaching the top with full preparations to occupy the same during November 1998. But the Higher Authorities ordered us to abandon the same. The Brigade Commander, Brig. P.C. Katoch, who had given me the orders was on leave at Delhi on that day and was not reachable that night. Our patrol could not have held on to the position without some kind of skirmish with the Pakistanis after morning first light. The Brigade Major conveyed the decision of the authorities just prior to the first light to abandon the mission. Pakistani forces occupied Thang Top during the war in 1999, giving them deep observation and surveillance inside the Shyok valley to our side. Pakistanis were continuously attempting to occupy many such heights along the Ladakh Range as well in the glaciers. It was a race for time.

Shyamal Sinha was from the Kumaon Regiment and was an instructor at the High Altitude Warfare School in Gulmarg when he along with three other Ladakh Scouts boys was attached to my unit to assist in rope-fixing during the operations. He went beyond the call of duty and volunteered to lead one of the assaulting platoons while ensuring tactical rope fixing for the attack. He was awarded the Vir Chakra. He later died in a mountaineering expedition in the Himalayas. Maj. Cheema was awarded a Sena Medal for this action. Any unit which has won a battle of this ferocity looks forward to the most prized reward of a 'Battle Honour'. However, my unit 27 Rajput appeared in the list of rejected claims for the honour in 2005. All 13 Infantry units awarded the 'Battle Honour, Kargil' were from the 8 Mountain Division and none from 3 Infantry Division which included my unit. I was told by the authorities that the honour was not bestowed on the unit due to our being deployed in the OP Meghdoot sector (Siachen Sector)! War is war. Unit valour in the same war should not have been segregated by some staff officers who were at least

2,000 km away from the front. The Staff Officers possibly advised the Commanders that being in Siachen Glacier Zone, our unit was not entitled to the honour for the Kargil War. I was also told later that the operations should involve at least two rifle companies for award of the same. Any right-thinking infantry man will tell you that it takes more than a company sized force to even support the operations of a platoon at such dizzy heights. The terrain simply did not permit deployment of more forces for the attack nor was it necessary.

Many years later in 2017, I invited Gen. V.P. Malik (Retd), former COAS, at the Infantry House in Mhow to join me for a drink which he readily accepted. He knew me as one of the Commanding Officers of one of the units which fought in the area north of Zojila Pass during the Kargil war. During the course of our conversation about the war I inquired from him if he was aware about the details of our action on Point 5770. He accepted the fact that he was not fully in the picture about the actual details of the operations. This indifference of the staff at various levels of the headquarters resulted in the unit and some individuals like Major Cheema not being recognised appropriately. I had recommended Major Cheema and Captain Shyamal Sinha for the highest gallantry award i.e. Param Vir Chakra. However, Major Cheema was awarded the Sena Medal and Captain Sinha was awarded the Vir Chakra. I had no words when the then COAS himself conveyed his regret of the same 20 years later to me. I requested him if I was permitted to quote his thoughts about the attack in written form to be filed in the unit war diary. He readily permitted me to do so. Accordingly, I wrote a DO (Demi Official) letter to Gen. Malik quoting our conversation with a copy to the unit to be inserted in the unit war diary. I, as an ex-Commanding Officer of the unit which fought the war, could not have asked for more 19 years after the war from the wartime Chief of Army Staff. It is surprising that the staff officers in the Military Operations Branch or in the lower headquarters could be so callous in reporting the progress of the operations to the authorities.

The success of my unit action had been prominently flashed in the local media as well as the national media. Messages poured in from my friends and colleagues. It was an inspiring moment of a lifetime that our countrymen applaud you just for doing one's duty. I recall hundreds of letters written by the school children from all over the country, mostly from Maharashtra state, with motivating words for our troops during the Kargil war. I got all these letters displayed for motivation of the troops. The song *Aye mere watan ke logon…* was truly one of the most inspiring words for us. No less was the contribution of the media and people from various walks of life. India became one. It is my conviction that wars can only be won if the nation stands together and supports the armed forces.

Our operation was a unique one in the entire Kargil war for many reasons. The attack was launched during daylight. No artillery was used before and during the attack though the Pakistani counter-attack was accompanied by very heavy barrage of artillery fire. Ours was the first attack wherein the involvement and presence of Pakistani regulars in the war was finally proven by producing the Identity Cards of their dead officer and soldiers.

In his thesis, 'High Altitude Warfare: The Kargil Conflict and the Future', (June 2003, p. 53) Marcus P. Acosta of Naval Postgraduate School, California has listed the capture of Tololing Hill by 2 RAJRIF (Rajputana Rifles) on 20 June, victory at Tiger Hill by 18 GRENADIERS on 8 July and the capture of Point 5770 (Balal Post) by 27 RAJPUT on 27 June as the turning points of the war. Yet, I still regret to this day that the highest award was denied to the unit on the grounds that the battle was beyond a line marked by the staff as the area of Kargil war! Be that as it may, I remain ever proud of my officers and troops and the fact that we played our part well in the war.

In the following days, other units in the formation under Brig. Katoch were also tasked to recapture the areas occupied by the intruding Pakistani troops in the north of Chorbat La Pass. This pass was the inter formation boundary between 3 Infantry Division and 8 Mountain Division which was inducted from the Kashmir Valley and deployed in the areas north of Zojila Pass and south of Chorbat La Pass. Massive firepower was used in other actions in the sector. There were other impressive successes in the sector. Both sides suffered heavy casualties. In one of the actions in south of Shyok River, two jawans of one of our sister units were also captured alive by the enemy. These soldiers were prominently displayed on Pakistani TV during the war.

A saga of bravery of this kind is unbelievable to many. Many could not believe it. At the end what mattered was our victory. No amount of recognition or kudos can replace the joy of my unit of having done its job against all odds. In retrospect, as a veteran of the war I feel that our troops are the finest in the world. Our officers in the field are the best too. But somewhere we lacked the structure and rules of business in the system that resulted in the war in the first place. This aspect has been adequately brought out in the KRC (Kargil Review Committee) report. The majority of the recommendations of the KRC are yet to be implemented. Many years have passed since the KRC was adopted. In fact it may be time to review the KRC itself and explore fresh ideas for the modernization of the Armed forces to meet the emerging challenges of the future.

By the end of the war we had operated in the entire length of the mountains north of Chorbat La to the NJ9842 point and Chulung Glacier. I often wondered what were the real aims of General Pervez Musharraf in the areas north of Chorbat La. Could he have been planning beyond occupation of the Ladakh ranges, and to take back the entire Pushto-speaking belt of Shyok (Turtuk/Chalunka) Valley? Or was he attempting to take a big chunk of the Siachen group of glaciers at the southern end of the Siachen Glacier? At every possible opportunity since the Indian Army's occupation of the Glacier Gen. Musharraf was part of the group that attempted to attack our posts in the Siachen area. The Pakistani army's occupation of Point 5770 and Chorbat La Pass, their occupation of Turtuk Heights, and subversion and training of some of the population of Turtuk village during the Kargil war were tactically brilliant manoeuvers which could have only been with the aim of occupying the entire Shyok valley and the southern edge of the Siachen group of glaciers. If they had been successful in doing so, this could have been their launching pad for further progress towards Nubra valley. Its outcome might have been disastrous. The only launching pad for the Indian Army towards the Northern Areas of POK was from the Turtuk sector. The Northern Areas, including the Pakistani army garrisons of Siari, Skardu and Gilgit, are in proximity to the Central Asian Republics and have close ethnic ties with each other. Should India have lost the Turtuk Sector, it would have been disastrous for us in terms of our reach and logistics towards the Northern Areas of POK as well as the Xinjiang region of China and Shaksgam valley in POK which is under Chinese occupation.

In his article 'Pakistan's General Problem', Pakistani author Mohammad Hanif wrote:

> In 1999, two days after the Pakistani army embarked on its Kargil misadventure, Lt Gen. Mahmud Ahmed gave a 'crisp and to the point' briefing to a group of senior Army and Air Force officers. Air Commodore Kaiser Tufail, who attended the meeting, later wrote that they were told that it was nothing more than a defensive manoeuvre and the Indian Air Force will not get involved at any stage. 'Come October, we shall walk into Siachen—to mop up the dead bodies of hundreds of Indians left hungry, out in the cold,'

General Mahmud told the meeting.[2]

Thus it was clear that Gen. Musharraf did have his eyes on the Siachen area, particularly the sector where I was. On the other hand, the entire focus of the war on our side was towards the Leh-Kargil road. The courage of my troops and other units in the re-capture of Chorbat La, Point 5770 near NJ9842 and Point 5810 in the Ladakh Range unsettled the grand plans of the Pakistani army to regain Turtuk-Chalunka and the Siachen Glacier in this sector. The national TV crews transmitted the battles near the Drass area since the live visuals of the bombings could be captured from the highway itself. It was the first of its kind in India that parts of the fighting could be seen live in people's homes. Heroic deeds of our soldiers and officers reached Indian households. The troops across the Khardung La Pass were fighting equally intense battles mostly unknown even to some of our own commanders. The units of the Trishul Division in the Shyok Valley area recaptured and threw out almost the entire enemy troops who had intruded in the area. At the end of the war, the performance of the Trishul Division units was largely overshadowed by that of the 8 Mountain Division in the minds of the hierarchy as well as the Indian public who never saw or heard of their battles.

In the year 2010, I was invited to be part of the Kargil Vijay Diwas celebrations at Drass. I went from Rajouri where I was commanding an army division in the South of Pir Panjal, proceeded to Drass via the Mughal route across Pir Panjal, Srinagar, Zojila and to Drass by road. I was taken aback when I didn't see many of the names of the units and personnel of the Trishul Division who fought and sacrificed their lives in the war memorial. I did take up the matter with the authorities.

---

2. Mohammad Hanif, 'Pakistan's General Problem', *Open Magazine*, 9 June 2011. https://openthemagazine.com/features/world/pakistans-general-problem/

The Siachen Soldier

At the end of the war we remained in the South Glacier for another one and a half years. The Siachen command tenure 1998-2000, including the Kargil war, was the most satisfying in my professional life. Troops were extremely motivated, particularly after a grand victory in battle. One can only 'feel' the Siachen Glacier experience. It will never be a complete story to narrate to anyone who has not been there in person. The intensity of mortar and artillery exchanges almost on a daily basis was almost comparable to a battle in an all-out war. Both sides used missiles, rockets, artillery and other small arms to dominate the other side. Both sides suffered casualties. In addition, the effects of the extreme high altitude, avalanches, crevasses of massive size, and the effects of the glaciated terrain can cause incalculable harm to the human body. No quantified measure exists as to the extent of harm to the human body in such an environment. It is a huge blow to the normal health of a soldier at such heights. I can certainly state from my personal experience that the physiological impact of the extreme high altitude and the glaciated conditions is indeed very severe. In a year or so after deinduction from the Siachen Glacier during 2001 while serving at the National Defence Academy, I started sweating only on one side of the body. My eyes started becoming drier and drier each subsequent year. The bones certainly became more brittle too. However, none of these were allowed to become any disability to me to perform my duties and I remained FIT till I superannuated 18 years later.

Physiological effects apart, the logistics issues were extremely challenging. Any fresh item, vegetable or fruit, took over 6-10 days on foot to reach the deployed troops from the roadhead. We survived mostly on tinned food items. There were a few posts which were cut off from the outside world for 5-6 months at a stretch. No one could go to these posts nor could anyone from the posts get out for these many months. Their needs were stocked up well in advance to cater to their survival as well as to fight the enemy. In one such post, the unit that we relieved in the area had an unfortunate fatal casualty. The body of the dead jawan could be retrieved only after four months. The jawan had died of high altitude sickness. We had Pisti, a high-altitude country dog, operating in the same post ferrying medicines and personal letters, etc. to the troops deployed there. The Pakistani troops were after her life. They frequently used medium machine guns at the sight of the dog to eliminate her. Pisti's companion Pista lost his life in one of the Pakistani firings at the duo. All my troops took special care to ensure that the Pista and Pisti were fed with their favourite food, a milk tin every day from their share of rations. At the end of the tenure, I recommended Pisti for a Commendation from the Army Chief. When I left Pisti finally, I took out one of the three commendations that I wore to pin it on her. I have been wearing one less in my uniform ever since. Finally, both of them received the Army Commander's Commendation in August 2000.

Our movements in the glacier were restricted to full dark nights only in most of the posts to conceal our movements. Such dark nights were available only a few days in a month. Every post commander had a chart of such days to plan movements. Helicopters were used to supply provisions only in places where helipads could be improvised and out of enemy missile range. Each helicopter could barely carry one or two jerricans of kerosene in every sortie. Most of the areas were prone to enemy missile attacks on our helicopters. Our pilots took considerable risk in flying to evacuate casualties.

I developed very high blood pressure during a visit to a post which was located beyond 18,000 feet. Many attempts by the helicopter pilots to evacuate me over a period of three days failed due to weather conditions. I decided to 'lose height' by walking down to the nearest helipad which was at least 12 hours walk. Accompanied by the doctor, we were able to reach the helipad after a night's walk. I was evacuated to the hospital where I was declared as having suffered from high-altitude pulmonary oedema (a potentially life-threatening condition involving abnormal buildup of fluid in the lungs). After two days in the hospital, I resumed my duties. Fetching ice blocks from nearby sources to melt into water for own consumption was also a challenge since it was not easy to get uncontaminated ice blocks. Human waste and gunpowder hardly decay in these climatic conditions. Coupled with these, the amount of waste deposited in the glacier is bound to have serious ecological consequences in times to come. One can easily estimate that close to five crore of non-biodegradable tetra-packs of milk, soft drinks and other such packages of items would have been dumped in these glaciers till date. This is in addition to hundreds of thousands of packets of other commodities supplied to the troops to operate in these glaciers. Despite logistic challenges, we initiated a cleanliness drive to dispose of maximum possible waste out of the glaciers through the returning porters during my command tenure there. More than 1,700 parachutes which were used to drop supplies were recycled. I recall that a similar exercise was carried out in 1987 after the Kanchenjunga expedition when we cleaned the area of Base Camp near the Green Lake in Sikkim and carried the trash from there for proper disposal.

Many posts in Siachen were constantly under enemy mortar and artillery attacks for most part of the day. We reciprocated these with more devastating firepower. Troops had to spend most of their time in ice caves for protection from the artillery shells and other types of missiles. During my visits to these posts I also started enjoying sleeping inside the ice caves. Ice caves provide the safest place against artillery shells. Ice caves are carved out of harder than rock ice available at such heights. It takes months to prepare a cave which can shelter 4-6 soldiers at a time. Psychological isolation of extreme kind is overcome by humour, camaraderie and good leadership at all levels. On 12 December 2019, I read a newspaper report about 74 army men having been killed in three years (2016-19) due to avalanche alone in J&K. It is very unfortunate that such incidents continue even today. Siachen is very prone to avalanches. We had to be very careful to avoid casualties due to such avalanches. Controlled movement was one measure to avoid casualties. There is not much we can do about nature.

I recall that I was away from home for more than a year continuously from April 1999 to June 2000. No leave could be planned due to the continuous flow of inputs of threats from the Pakistani army. There was limited access to a satellite phone at subsidised rates to call home once in a while. This facility was put in place in 1998, one phone for about 2,000 soldiers. The DRDO (Defence Research and Development Organisation) detachment nearby of 8-10 people for avalanche warning had two of them!

I learnt much later that my house in the village was subjected to searches by the local Assam Rifles troops during combing operations in my village searching for the insurgents. My aged father and two-year-old niece were taken out of their beds in the middle of the night. I pacified my family that the soldiers were doing their duty and that some inimical informers must have given wrong and motivated information to the troops located at Thoubal. This was the time when I was busy fighting the Pakistanis during the Kargil War. At the end of our tenure in the glaciers, we came out with flying colors having gone through the vagaries of nature and the fight with the enemy. Our job was to complete an assigned task at the unit level.

Many veterans and civilian friends asked me if all these efforts and the costs involved in the Siachen was worth it at all. It cost the country Rs 3 crore a day in 1998 to maintain the forces there. My answer to this aspect has been that it was a part of nation-building efforts of the 'Idea of India' and that must be upheld at whatever the cost. The expenses could be compared to insurance premiums for the country. I was surprised years later that some backchannel efforts by highly connected ignorant persons from India had suggested a mutual withdrawal from the Siachen Glacier by both the Glacier proper countries. It was a laughable idea indeed. The Pakistani army were nowhere in the vicinity of the Siachen. They were only in a few locations of the Saltoro Range which overlooks the glacier from a remote distance. Their lines of military communication were a lot shorter and much easier for military buildup compared to ours. In the backdrop of the Kargil intrusion and the reliability quotient of the enemy on the agreements signed between the two countries in the past, it was unwise to suggest such a mechanism to the decision makers by an ignorant and inexperienced few who claim to be experts on the Siachen issue. To quote Lt Gen. K. T. Parnaik, former Indian Army Commander in J&K, who stated during his interview by Nitin Gokhale, 'Don't forget, Kargil happened because of Siachen.... If you peruse their own records, which are now in public domain, one of the major objectives was to force us to vacate Siachen.' (*Beyond NJ 9842: The Siachen Saga;* p. 263). I am of the opinion that the issue of Siachen is an issue of boundary between India and Pakistan. Premier Benazir Bhutto of Pakistan, in a complete reversal of her earlier stance, declared in 1994 that the resolution of Siachen is inextricably linked to the larger Kashmir problem. Maj. Gen. Afsir Karim, a renowned Indian defence analyst, in his article 'Pakistan's Aggression in Kashmir 1999' in *Aakrosh* (July 1999, Vol. 2, No.4, p.13) writes:

> The root cause of the conflict is no longer Kashmir, but the sway of fundamentalism in Pakistan...Talibanisation and chaotic conditions within Pakistan can make the danger of a nuclear war quite real in the region...... Monolithic domination by fanatics of any variety in the strategically sensitive area of Afghanistan-

> Pakistan, Central Asia-Kashmir conjures up a frightening scenario.

The costs of the conflict on both sides of the Siachen were far heavier before 2003 than it is now, since there is no real fighting today. The responsibility of the military is to ensure integrity and security of the nation. The issue is about national interest and national security. Talks for resolution broke off repeatedly on elementary military issues like whether a disengagement of troops be called 'withdrawal' or 'redeployment'. How to ensure that a demilitarized zone remains inviolate. Military commanders of both the countries started the war with little political reason. The politicians later used the conflict to draw mileage. A setback in Siachen for Pakistan would have cost heavy political price in later years.

On completion of our unit tenure in the Siachen Glacier, we moved to Rajasthan for a peace tenure of the unit. I felt satisfied that our troops had done the regiment and the army proud during the Kargil war as well as in the highest battlefield of the world. There was very strict checking of the de-inducting unit against carriage of any unauthorized items from the glacier. I had ordered that a few items like war souvenir, discarded climbing equipment and some other such stores be carried to be kept in the motivation room in the next location. Some special rations like chocolates and dry fruits which were authorised to the troops but not consumed due to the late arrival of these stores were also to be carried along. It was learnt later that some staff officer in the higher headquarters had ordered an inquiry against the transportation of these items. It was a shocking application of mind and rules. I recall that I had given in writing that it was I who ordered the move of these items and that I was responsible for the same and action if any should be taken against me and no one else. Nothing was heard thereafter!

The Siachen guns became silent in the Highest Battlefield sometime in 2003 after an agreement between the two neighbours. I am happy that Siachen remains free from booming guns for the last 18 years. It has been over 20 years that I last set foot in the Siachen. I still miss the ecstatic feeling of just inhaling oxygen in lower altitudes after descending from the super high altitudes.

As a student of the Sainik School in Goalpara, Assam, in 1968

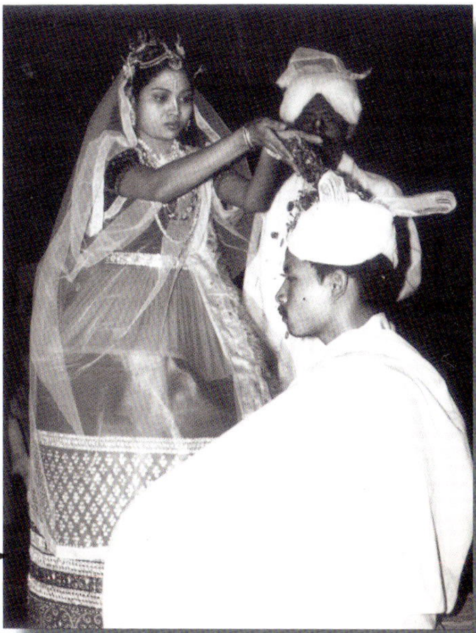

During my marriage with Capt. (Dr) Mangala at my village Charangpat, Manipur, on 18 November 1983

Atop the northeast spur of Mount Kanchenjunga during the Assam Rifles' Expedition in 1987 (Note the tiny camp tucked amid the ice in the right side bottom corner)

Briefing Prime Minister Rajiv Gandhi about the Assam Rifles' Expedition to Mount Kanchenjunga at his official residence in New Delhi in 1987

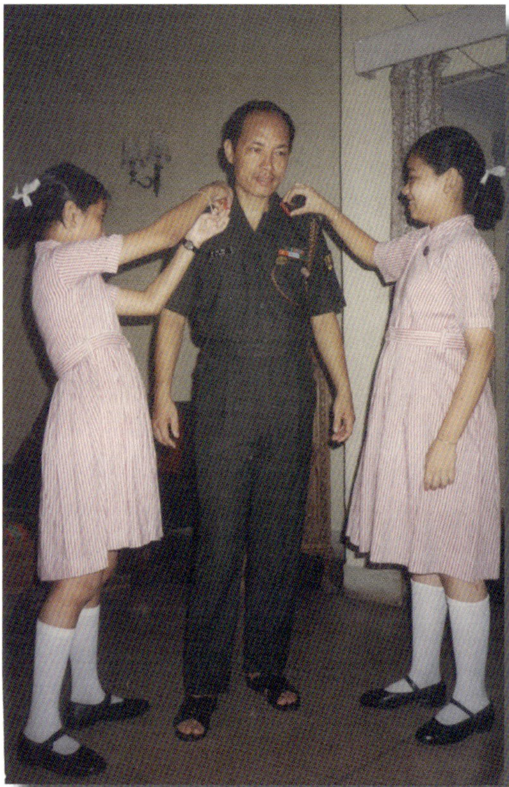

My daughters Poornima and Priyanka pipping the Colonel's rank on my uniform in Delhi in 1997

Posing with a double-humped Bactrian camel in Nubra Valley in 1998

Enjoying river rafting with my unit members in the gushing waters of Shyok River in Nubra Valley in May 1999, prior to the Kargil War

Major Navdeep Cheema, SM, (now Colonel) who led the assault team to capture Point 5770 during the 1999 Kargil War. Point 5770 is now known as Navdeep Top

Picture of Point 5770 taken by me from the glaciers before its capture

At a post in Siachen Glacier during the 1999 Kargil War

Honouring Pisti, a high-altitude country dog, with one of my Commendations. She helped us keep stress at bay in Siachen

Patrolling the AGPL in Siachen in 1999 on a snow scooter

View of Zanskar Range from Digar La Pass in Ladakh Range

An unidentified missile that struck near my bunker during the 1999 Kargil War

Holiday time with family in Diskit village, Nubra Valley, after the 1999 Kargil War

With Chinese officers at Nathu La Pass, Sikkim, in 2004

Calling on former President of India Dr A.P.J. Abdul Kalam at his residence in New Delhi in 2007

En route to Drass with my wife Mangala and daughter Priyanka to be part of the Kargil Vijay Diwas 2010 celebrations

Interacting with the *Awam* (local villagers) in Kirni village, Poonch district, in J&K in 2011

During a briefing with one of the Commanding Officers in 2011 after taking charge as GOC, 25 Infantry Division, in J&K

With Indian cricket captain and Lt Col (Hon.) Mahendra Singh Dhoni during his visit to Rajouri in 2011

Talking to soldiers of my Division in 2012

With Defence Minister Shri Manohar Parrikar during his visit to J&K in 2015, where I was the Corps Commander

With my brothers, their families and my close personal staff during a get-together in Nagrota in 2015

My wife Mangala and me at the
National Defence Academy campus in
Khadakwasla, Pune, on 16 December 2016

Paying homage to the martyrs of 27 RAJPUT before hanging up the uniform in February 2017

Receiving the Param Vishisht Seva Medal from Hon'ble President of India Shri Pranab Mukherjee at the Rashtrapati Bhawan in New Delhi on 20 March 2017

With Governor of Manipur Smt Najma Heptulla after delivering a lecture at the Raj Bhavan, Imphal, on 25 May 2017

Mangala with our eldest daughter Poornima, her husband and in-laws in San Francisco, USA, in 2018

With Mangala, Priyanka and her husband Vendru on my last day as Chairman of the Manipur Public Service Commission, in February 2019

After a lecture at the National Defence College, New Delhi, in March 2019

4

Soldiering On: Towards Ethical Officership

The Cradle for Leadership

On completion of three years and four months in command of the unit, I joined the National Defence Academy (NDA) at Khadakwasla as a Battalion Commander in 2001. My job was to oversee the training of about 400 cadets of the total 1,500 cadets under training at any time at the NDA. These cadets were being trained to become officers in the Army, Navy and Air Force of the country. I left 27 Rajput unit with a very heavy heart and bid adieu after handing over the command to my able Second in Command Col S. Bhattacharya. I enjoyed every moment of command of *Sattaees*. It was a journey most memorable and one of the sweetest songs in my life.

At the NDA, I observed that there was a further need to develop ethical and inspirational leadership among the future leaders. A need was felt to widen the horizon and add intellectual depth to the training curricula of the cadets. The days of phrases like 'Beg borrow or steal' or even 'End Justifies the Means', 'Survival of the fittest' which were the buzzwords in the past needed to be discarded. My mind went back to the unfortunate death of a cadet in 1975 when the senior cadet was ragged to death by juniors during the squadron social at the end of term. The dropout rate during the initial days of training was considerably high i.e. upto 10-13 per cent when I joined the NDA as the Battalion Commander. The challenges of making future leaders for the 21st century are many. The existing training curricula called for finding ways on the future needs of the Armed Forces. The present academic system mainly focused on the award of a Bachelor's degree.

Introduction to subjects such as Military Psychology, Management, Space Science, Cyber Warfare was the need of the hour. The quality of the newly recruited teachers was found to be less than of desired standards. The service conditions of the teachers had been gradually reduced, hence the downward trend in the quality of teachers. Creativity among the cadets was discouraged by the senior cadets. The late Major Jaswant Singh (who was External Affairs Minister during the Vajpayee Government) is quoted as having said in an informal gathering that, 'NDA destroyed my soul and free spirit' and that he somehow 'retained sanity through reading and music'. All this and much more of what I saw as cadet in 1974 appears to have been perpetuated in some measure. But the NDA of yore forgot that perpetuation of such juvenile sadistic tricks has profound negative impact on the character building of leaders of the future. The attitude among some serving officers of the day to lobby for favours germinates from such exposures during the formative years. I am convinced that leaders of tomorrow will need to possess qualities of strong character, unwavering will and uncompromising resolve to do good for the nation.

Military discipline is the key to all training activities for the newly inducted personnel. At the NDA and elsewhere in the Indian Army, drill parades form the bedrock of discipline. The best way to maintain discipline is to ensure that the person prides himself or herself in doing the things he or she ought to do. Discipline is also achieved by one's pride in the unit, one's belief system or the profession he/she is in. The worst form of maintaining discipline is by instilling fear. In the NDA, fear was the main instrument of maintaining discipline. My endeavour was to train and instil discipline by self-pride and pride in one's profession.

Good drill is about uplifting of the spirit of pride in the uniform and inspiration. But the pomp and show of drill should be for special purposes only. At the NDA, preparations for the passing out parade drill became a priority over other activities of training. I was of the opinion that this needs to be curbed. Drill learnt for the battlefield had long become history after the 'Oblique Order' of Frederick the Great in the Napoleonic war, where hundreds and thousands of soldiers lined up in battle formation as per prescribed drill. Today drill is still relevant to ensure orderly execution of military operations such as launching of a bridge by the Sappers, the gun drill by the Artillery, drill for mobilization for war and so on. The focus is on learning by practice to do something right so automatically and willingly that it becomes natural even in times of warlike conditions. Training and instilling discipline among the young cadets to preserve the initiative and independence needed in war was what I was looking at in the NDA. I looked at the officer cadre — as servants of the society, as leaders with character and above all, as war fighters.

I proposed and discussed my views with the then Commandant, Air Marshal A.K. Trikha, a very erudite officer. He agreed with my observations but he was only partially in favour of the many changes that I had recommended. He was of the opinion that changes should be further deliberated upon. He was also hesitant on the use of ancient Indian texts as case studies to impart education on moral and ethical officership. He felt that these efforts could be construed as religious content by some quarters. I tried to impart in the cadets such varied topics as the *Soldiers duty to the nation*, *Loyalty to the Constitution*, *Loyalty to the Government* or loyalty to the *commitment to defend the values of our nation*. According to me, the role of a commander or a leader is to be a tactician, a warrior, a strategist, a manager and a moral standard bearer. He must have insights and values associated with courage, temperance and justice. He must be in Plato's words a 'virtuous leader'. Towards this end, I made a serious attempt to introduce a Principled Approach to Officership among the cadets in my Battalion.

I began with deliberations on available case studies of officer-men relationships. I made it a point to be present myself as often as possible among the cadets. The cadets accompanied by me went cycling and camping outdoors many times. In these informal outdoor camps, we would discuss and debate case studies on leadership. I do not know how much impact it had, if any, on the cadets, but I am convinced that some of the old ways of so-called traditions in the NDA must go. I felt it necessary to address the attitude and the character qualities of officers posted at the NDA to train the cadets. The Officers Code at NDA was approved by the Commandant. It was meant to serve as a personal guide to the newly inducted officers in the Academy. It primarily appealed to the officers to imbibe a Principled Approach to Officership.

As far as the physical hardships during training are concerned, there is no denying the fact that cadets of the Academy must be trained to be able to withstand physical and mental pressure under difficult circumstances. To achieve the same, a certain degree of unplanned activities became necessary. However, avoidance of any kind of sadistic ragging or manhandling was the need of the day to sustain a good training environment.

During the turn of the century, it is estimated that more than 80 per cent of the cadets in the NDA were from middle class or lower middle class background. Around 20-30 per cent cadets were sons of Junior Commissioned Officers or other ranks and equivalent from all three services. More than 50 per cent cadets were from Sainik Schools or from Kendriya Vidyalayas. Another 35-40 per cent were from government schools. Therefore, it is obvious that sons of the rich and powerful, high ranking defence officers, and bureaucrats did not consider the armed forces as an attractive career option. The good part is that cadet material from the villages and low income groups tend to be hardier.

As I write this, we are witnessing a change in the moral compass among the officer cadre today. Our countrymen and women are proud of the Armed Forces. But in my opinion each one of the members of the Armed Forces needs to do more for the country. Military ethics need constant attention and practice by the senior commanders in particular. I recall a letter written by Gen. K. Sundarji as the COAS dated 1 February 1986, i.e. 34 years ago, to all army officers regarding the moral values in the army. He wrote: 'As a whole, the Corps of Officers has lost much of its self-esteem, pride and élan; it is becoming increasingly careerist, opportunistic and sycophantic; standards of integrity have fallen and honour and patriotism are becoming unfashionable.' I received the letter as a Captain while serving in Mizoram in 1984 and the contents are as relevant today as before.

As I rose in rank in the hierarchy later in service, my views on officership were reinforced further. The ethos of all ranks, particularly of middle ranking and senior officers, needs to be reoriented in keeping with the times and in keeping with the operational needs. We have had cases of 'Ketchup Colonel', 'Booze Brigadier' two decades back and others who have proved to be liabilities in the society leave alone in the army. Many cases of corruption have been detected, misuse of authority, sex-related offences, and above all many cases of practice of 'Self before Service' have been reported in the media. The debate over politicization of the army, never-ending debate on civil–military relationship, superior-subordinate relationships, lack of personal qualities, virtue and leadership are dominating the discourse among the military thinkers today. There is a need for renewal and redefinition about the traditional understanding of military professionalism. All these qualities need to be nurtured among the cadets undergoing training in the academies.

More of Defence Management

My tenure at the NDA was a short one as I was nominated to attend the one-year-long prestigious Defence Management Course at the College of Defence Management in Secunderabad during 2001-02. I was not expecting to be nominated for the prestigious course since the first command report I earned as a Commanding Officer in Delhi was not up to it. However, I later learnt that the Army Headquarters had given due weightage to the performance during the Kargil War and operations in Siachen, while drawing on the merit for the course. The course was a very educative defence management programme. I, however, felt that officers should have been exposed to a compressed practical version of the course curriculum preferably before command of a major unit. It also gave an opportunity for intellectual exercise of one's thinking powers. My dissertation for the award of a Master's degree in Management Sciences (MMS under the aegis of Osmania University) was 'Ethical Officership in Indian Army 2020'. The hypothesis of the paper was 'Ethical officership is the only way to lead and win wars of the future'. There are differing views on the subject. The findings of the research which was carried out in 2002 reflected the stark reality that we needed to do a lot more towards identifying and grooming the senior commanders in the organisation. I am proud that many of the senior officers I interact with today have it in them. The need is to make ethical officership a culture, just as the Americans attempted after the humiliating defeat in Vietnam.

On completion of the course I joined a Mountain Division in Sikkim as Colonel Administration to look after the logistics and manpower issues of the formation. My daughters had grown up and were pursuing their respective degrees in Pune. Priyanka preferred to pursue a degree in psychology and Poornima was in an engineering college. Mangala continued to work in the CGHS (Central Government Health Scheme), Pune. I remained away from the family due to the service conditions. Mangala was managing her job, home and the daughters with panache. She lived up to the expectations of an army wife. Most officers, particularly in the Infantry, face such separation more frequently. Army wives contribute immensely to the nation by taking full care of the family while the husbands are away on operational duties. Mangala was no less.

China Front: Sikkim

I had visited Sikkim earlier during the Kanchenjunga expedition 1987. The natural beauty of Sikkim is well known. The state is one of the most attractive tourist destinations in the country today. Nepalis, Lepchas and Bhutias co-exist in a peaceful manner unlike some other states in North East India. The erstwhile Kingdom of Sikkim was merged with the Indian Union after a referendum in 1972. Prior to its merger, the Sikkimese kingdom had a long history of association with Lhasa in Tibet and relationship with Bhutan and Nepal as well. The Britishers treated the Kingdom of Sikkim as a strategic asset. The boundary between Tibet and Sikkim is clearly defined but border issues do crop up often between the Chinese Army and the Indian Army, particularly in the area of Doklam and the northern high altitude plateau. It has a significant strategic importance for India.

The 20-km wide Chicken's Neck or the Siliguri Corridor lies to the south of the state. Nathula Pass in east Sikkim connects with Tibet in China for trade. The Doklam standoff incident of 2017 between the Chinese forces and Indian Army reflects the significance of the area to our security. Skirmishes between India and China have taken place in Nathula during 1968 over the boundary issue.

China's brutal assault on Indian troops on 15 June 2020 at Galwan valley in Ladakh is a classic example of Sun Tzu's principle of warfighting, 'Kill one, terrify ten thousand'. It is clear that China is not willing to share the Asian strategic space with other Asian countries even while asserting to the United States that the Pacific has adequate space for both US and China! The geopolitical calculus of the 21st century in every part of the world has been severely altered by the Coronavirus pandemic. In the emerging new great game of the 21st century, the Sino-India military face-off in Ladakh has added further new dimensions, in the manner that India is emerging as a 'Swing State' against China in the new world order. It is in this context that the 'Tibet Card' could emerge as a very attractive option for India, just as China plays the 'Pakistan Card', and 'Arunachal Card'. It is a matter of time before the Indian leadership sheds the policy of underplaying the 'China Threat' and asserts India's sovereignty over its territory. It is a stark reality that China, in the emerging world order with its military infrastructure in Tibet, is likely to look at parts of Ladakh, Sikkim and Arunachal (the areas China considers as part of the five fingers of Tibet along with Bhutan and parts of Uttarakhand) to 'teach India a lesson', as reports in some Chinese military literature suggest.

China and India have no differences over the boundary issue in Sikkim itself. But, the Chinese forces continue to perpetuate tensions in parts of Sikkim, such as in Nakula area in North Sikkim besides Doklam in the trijunction of India, Tibet and Bhutan. Chinese forces had shown their intention to grab the sensitive area of Doklam even during 2003-2004, during my posting in Sikkim. However, we were able to match their actions and managed to keep the status quo even then. The Chinese were firmly stopped at the place in Doklam where we wanted them to, once we made up our mind. There was no elaborate social media to bring all those in public domain for TV debates then. China's claims over areas in Bhutan which could seriously jeopardise the security of India in the Siliguri corridor, to the south of Sikkim, is of much concern for the security of the North Eastern part of the country.

It was during 1986, while preparing to climb Mount Chomoyomo near the North Sikkim plateau region as a trial run for the main expedition to Mount Kanchenjunga in 1987, that I saw very minimal Chinese army movement in the form of vehicle mounted patrols across the International Boundary. No aggressive intent was visible during the period. But by 1987, General Sundarji, the COAS, had launched 'Operation Falcon' in the watershed of the borders in the Himalayas in response to Chinese movements across the entire eastern borders. I met many Chinese PLA (People's Liberation Army) officers during various formal meetings in 2003-05, which were normally held at Nathula. The Chinese believe in psychological domination over any potential adversary. Therefore, they try to attempt aggressive behaviour to read the minds of the others. I also suspect that they have a false sense of civilizational and ethnic superiority over other nations. As a professional soldier, my reading of the PLA officers is that they have a narrow military vision in terms of the Communist Party teachings and poor tactical acumen. During the Deng Xiaoping era, the Chinese significantly reduced their involvement in insurgency-related activities in India's North East, though some support through 'deniable activities' is believed to continue till today. They appear to have a false assessment that people of the region will support them against India in the long run. This assumption is totally false. In fact, the Chinese treatment of the Tibetans, the Uighurs in their own territory and their treatment of Myanmar, Vietnam and other countries are too well documented to be glossed over by anyone in the region.

Given the unstable nature of polity in the region and insurgencies in some of the states, it can safely be assumed that the Chinese will certainly attempt a 1962 in the region in the event of a future military confrontation, least realising that India has gone past the mindset of 1962 and is in a position to thwart any such adventure from the north of the Himalayas.

The Chinese forces are most likely to continue to keep the pot boiling in the Sikkim and Ladakh sectors to leverage Sino-Indian geopolitics in the future. I am convinced that China will not let any part of the LAC (Line of Actual Control) remain peaceful in the foreseeable future. Therefore, our forces need to develop capability to match theirs at all times.

The Sena Bhawan

After two years in the assignment in Sikkim, I joined the Army Headquarters as Director of Personnel Services. Mangala joined me in Delhi once again while the daughters continued for some time in Pune to complete their degrees. We were lucky to be able to arrange a transfer for Mangala to Delhi. The tenure in the personnel branch was a very challenging one. My job was to primarily deal with the Terms and Conditions of Service of all officers, men and women, in the army. The job involved much closer and frequent interaction with the Ministry. Implementation of cadre review for officers, new policies for Short Service Commission officers, gender equality of men and women in the Defence Forces, Sachar Committee issues (a panel set up to examine the socio-economic and educational status of the Muslim community in India), Pay Commission issues, were some of the important matters at hand when I took over the assignment. I discovered that most of the senior officers of the rank of Brigadier and above, leave alone other officers, had much to learn about matters related to human resources. Most of the subordinate headquarters at Command or Corps level and below, did not even respond to our correspondence with them to seek their comments on draft personnel policies. The attitude of negligence or even ignorance about personnel matters among the commanders and staff was reflected in the much-debated degrading status and spirit of the army over the years. I made my own priority among the many pending issues, to pursue what I thought were the important areas for the army.

The approval and promulgation of the new policies for Short Service Commission for men and women were completed in 2006. For this I must mention the clarity of thought and understanding of the issue by the late Shri Pranab Mukherjee, who was the Raksha Mantri then. He was solely responsible for ensuring that terms and conditions of service between men and women Short Service Commission Officers in the army had near parity. The response of the army to these suggestions at that time was that the women officers needed to be given adequate time and opportunity to be absorbed in the system. The implications of giving Permanent Commission to the women officers particularly in Combat Support Branches such as Corps of Engineers or Corps of Signals were many.

Recently, the Government of India announced that Permanent Commission was being given to women officers in all the branches except the Combat Arms i.e. Infantry and Armoured Corps and also the Artillery. It implies, therefore, that women officers in the Corps of Engineers and Signals, subject to conditions, were theoretically eligible for the General Cadre who command large body of troops and formations in war/operations as also to rise to be the Chief of the Army Staff. During the debates on TV channels, some analysts highlighted achievements such as command of a contingent in a drill parade, posting as defence attaché, and also drew comparisons with the women in the Central Armed Police Forces (CAPF) to support the cause of command assignments for women officers. Make no mistake, the role of the military is far different from the Central Armed Police Forces. The Directorate of Public Relations, Ministry of Defence, also circulated articles on 'Women in Armed Forces' to sensitise the environment on the issue. The business of fighting in our context is little known to many outside. The visuals of golf and glory, sparkling cantonments and parades which show peacetime dimensions of soldiering in the media are far from the ground realities of death and destruction. I am confident that the women officers will do well in command of Administrative and Support Services units. It remains to be seen how the women officers are deployed in the coming years in the best manner. I am sanguine that with more experience in the environment, the women officers should perform well in other fields.

I have attended two interviews for UN assignments during my army Services for postings abroad. The second such interview for a posting to Lebanon was held a year prior to my promotion to the rank of Brigadier. I was not selected for the assignment this time too. In times like this one tends to realise one's own deficiencies in the profession since it was the second time that I was rejected for a foreign posting. Military environment is a fairly competitive one. Rejections in any selection process whether for promotions or nominations for better assignments do lower one's confidence in oneself. I knew that I had to improve my communication skills far more.

Manipur Gets the First Brigadier

In 2006, my childhood dream materialised! I was promoted to the rank of Brigadier, thus becoming the first Manipuri to become a Brigadier in the Indian Army. Indeed, the first One Star Rank from Manipur in all the three branches of the Defence Forces. It became a matter of pride for the people of the state that Manipur finally had a Brigadier. Other states like Nagaland, Mizoram, and a few others in North East India barring Assam had not produced any such high ranking officer in the defence forces. Many congratulatory messages poured in from various quarters. The late Shri Rishang Keishing, one of the longest serving parliamentarians and ex-Chief Minister of Manipur, held a special meeting for the Manipuris to felicitate me at Delhi. I owe this humble achievement to my parents first and foremost. I am indebted to my parents, teachers, and my elders who guided me and supported me in every way. My family that has steadfastly stood by me deserves a special place.

One of the most memorable congratulatory messages that I received was from Lt Gen. V.G. Patankar (former Corps Commander in Srinagar) who was my boss when I was in a Counter-Insurgency Brigade in Manipur during 1992-93. He sent me an article titled 'How to Manage One's Boss' along with a letter stating that he was convinced that I was poor at managing bosses so I needed to learn to manage my boss better now! How true! During my entire career, most bosses have expressed their displeasure at my inability to manage bosses. I was lucky to have realised my dream, a dream which I had penned down in my diary as a child. This achievement was not only a dream realised in my personal capacity but for the thousands of young people of the state who were yearning to see a 'Brigadier' from amongst one of their own since India's Independence.

I often wonder how our predecessors from Manipur or any other North Eastern state did not reach higher ranks. I estimate that there have been more than 20-25 Manipuri officers who preceded me in the defence forces in the officer cadre. Was it due to lack of requisite educational and military qualifications, or battle experience on their part? Or was it due to lack of personal qualities expected of a senior commander? Or was it due to lack of hard work? Most of my older colleagues from North East states in the defence forces are well read, they were position holders in their school and even during military academy days. What could have gone wrong along the way, and why did only a handful of them go on to even command units leave alone rise to the rank of Brigadier or equivalent and beyond? Perhaps it could be more due to the lack of additional qualifications such as being a Staff College graduate, not clearing the promotion examinations in time, or even lack of communicative abilities, and even attitude towards the idea of soldiering itself. Being relatively poor in Hindi language and of particular accent, many officers from North East India including myself were at a disadvantageous position too. Most were individually very brave officers who perhaps could not get the opportunity to display the same when the time came to do so. It is a very competitive environment in the higher ranks in the defence forces. Did some of them burn out early?

One has to be better than the others professionally in every aspect. One is also expected to be well read, communicate well, have a wider vision, have intellectual depth, possess most of the important leadership qualities in the true sense, and above all some luck in war or war-like operations to come out with flying colours. I am hopeful that more of the future generation of Manipuri officers in the defence forces will reach the star ranks in the not so distant future. As I write today, I am happy that my ex-Colonel Military Secretary, Col N. Romeo Singh of my unit and of my state became the fourth Brigadier in March 2020. There are over 300 defence officers from the state today, thus making Manipur have one of the highest per capita number of defence officers in the country. Many of them have won accolades in the army, been awarded very high gallantry awards. The culture of doing well is catching up. No amount of blaming the system for failure to rise in the rank will be in order today. The glass ceiling has been broken. I am confident that many more from the state and other parts of the North

East can do it.

I was given the rare honour to command an Infantry Brigade stationed at the national capital. Earlier, I had requested the Additional Military Secretary, in charge of postings of Brigadiers, that I would only draw Re 1 as pay if I was allotted the Siachen Brigade to command! He stated that I could not get the formation I wanted. Command of a formation in the capital had its own challenges. I was lucky to have the best of the infantry battalions in my formation. It was part of one of the Strike Corps of the Indian Army. The concept of fast mobilisation and deployment in battle positions before the adversary could strike, popularly called 'Cold Start' doctrine, was conceptualised and introduced a few years earlier. The entire operational focus was on quicker mobilisation of the strike formations during the period. There had been no official document or a doctrine known as 'Cold Start' doctrine. This doctrine the supposedly introduced first after 'Operation Parakram' in 2001-02. Lessons learnt from 'Operation Brasstacks' 1987 and 'Operation Parakram' formed the basis to evolve the concept. I studied the concept and its application to the Infantry Formation which was part of the mechanised Formation. My exposure to Mechanised warfare so far was limited to theoretical discussions in various courses of instructions. I never had a hands-on experience. Therefore, it was important that I mastered the mechanics of the concept.

Many friends from the Mechanised Forces helped me in my endeavour to settle down early in the mechanised environment. My priority for the men I commanded was Training for War and to maintain a high level of morale and readiness. Commanders at all levels supported me in the endeavour. Administratively under Delhi area, one had to maintain constant liaison with the Area Headquarters to ensure that quality of life for the troops was maintained well. The command tenure as an Infantry Brigade Commander is a short one, normally 20-22 months. I attempted to inculcate the spirit of adventure among the units. Accordingly, all the units of my formation were motivated to plan and execute a high altitude trekking, a mountaineering expedition or similar activities. I truly believe that the best training of troops for war is war itself. War is too expensive a way to train the army. Therefore, any adventure which necessitates taking calculated risk of life and death is the next best way of training of leaders for war.

The role of a Brigade Commander in a station like Delhi is quite a challenging one. There is need to balance operational preparedness and administrative needs of the organisation while maintaining highest level of morale among the troops at all times. Contrary to what some may believe, the administrative needs of the army in the capital city from the Brigade was minimal. Commanding Officers were completely free to rest, refit and train their command, besides attending to their regimental and personal commitments. It was, however, not a challenging tenure in terms of operational assignment. Deprived of any operational achievements to show in a peace area, one often finds the senior commanders looking for administrative acumen or performances not directly related to operational needs. This trend among the seniors needs to be curbed. One need not be in the operational areas to judge the capability of an officer. In any case, not every command can be in operational areas for everyone since the command assignments in these areas are limited in number. I was also to discover that I needed to work much more towards gaining in-depth knowledge of commanding a formation in a mechanised warfare environment.

It was while commanding the Brigade that I realised that I was ageing! During a football match between officers and the JCOs of the brigade I observed that the players were moving too slowly in the field. I went inside the football field to play on the side of the JCOs. While my mind was fast enough to tackle the ball, my body was not quick enough to catch up with the ball at play. I tripped a few times and fell during the game. I became conscious of the fact that my physical fitness at 51 was slowing down and I needed to work on it.

A Special Guest: Dr Abdul Kalam

One of the memorable events during the period was my interaction with ex-President the late Shri A.P.J. Abdul Kalam while he was accommodated in a facility managed by my brigade. He had opted to stay in the Cantonment till completion of repair works of the house allotted to him after leaving Rashtrapati Bhawan. He called me a few times over a cup of tea to talk about North East India. He even asked me to go back to Manipur to settle down after my army service. He reasoned that indigenous people who have worked outside the state in higher responsibilities can bring in progressive changes to the state. I told him that there are many wise men in Manipur but they hesitate to display their wisdom. He understood and smiled. I was surprised that he knew many more issues about the state than I had imagined him to know. He observed that many at the helm of affairs in the state lacked the courage of conviction in them. He mentioned that only honest and dedicated people could bring about changes for the better. He opined that the future of the state appears bleak unless positive changes are brought in early. He was very concerned about the state of affairs in the North East in general and Manipur in particular. He foresaw the shape of things to come in the region and shared his opinion with me as early as 2007 in Delhi.

Fort William: No. 2 Rampart

As my tenure in the brigade ended in August 2008, I joined the staff of Lt Gen. V.K. Singh who took over as the Army Commander in Eastern Command based at Fort William, Kolkata. The No. 2 Rampart in

Fort William was a reserved accommodation for the assignment. The ramparts were the inner boundary ditch-cum-wall around the famous fort from where the British conquered and ruled India! The climb to the house was through a tunnel like structure on the ramparts of the fort. To find a beautiful elevated lawn after climbing the tunnel was the speciality of the houses there! The British East India Company had constructed tunnels from the fort to the River Hooghly a few kilometres away.

My job was to deal with the issues of officers related to complaints, Confidential Reports, honours and awards, postings and transfers, besides being the Senior Personal Staff to the General. The Honours and Awards Committee of the Command Headquarters chaired by the Chief of Staff sends the recommendations to the Army Commander. It was one of my jobs to scrutinise the same. I made efforts to ensure that only the most deserving cases were recommended. In one odd citation, there was mention of heavy exchange of firing with the militants and yet we found no trace of the effects of militants firing at the site of incident. There were allegations of fake encounters being conducted in Manipur and elsewhere. The case of Major Sreeram Kumar who was awarded the Ashoka Chakra in 2009 is a case in point. A panel investigating this particular case had opined that it 'was not an encounter but an operation by the Security Forces wherein death of the victims was caused knowingly'. (*Blood on my Hands* - Kishalay Bhattacharjee, Harper Collins. p. 26). Major Sreeram Kumar was later charged with 'fake encounter'. The fact remains that the officer was recommended and awarded the highest peacetime gallantry award of the Indian Republic. Obviously, some things have gone wrong and the guilty not been punished. It was important that such doubtful cases should have been rejected by the awards committee. These are only exceptions than the rule. The system is very fair. But, there is a need to ensure that the guilty are not rewarded.

the Defence Forces of our country are one of the best in the world. However, I also felt that the organisational structures, decision-making processes, civil-military relations, politico-military interface and research on security related matters have plenty of room for more keen thought and taking corrective measures. The earlier the better for the country. The learning process from peers gave us insight into their respective services. One year in Raksha Bhawan near India Gate in community living was most memorable for the families as well. Almost 80 per cent of the course participants rose to become Three Star Generals or equivalent in the armed forces while our civil counterparts also rose to similar positions. Even to this day, frequent meetings, bonhomie and interactions continue regularly.

On completion of the NDC course, I joined duty as the Deputy Military Secretary in the Military Secretary's Branch, in the Army Headquarters. I relieved my friend and coursemate Brig. (later COAS) Bipin Rawat who proceeded to command the Congo Brigade. My responsibility included dealing with Promotion Boards upto the rank of Colonel (later included upto the rank of Brigadier), complaints, commissioning in the army, release and retirement issues. No doubt, it was a challenging assignment. The posting at the Army Headquarters also gave me an opportunity to be with my family in Delhi. I had the opportunity to be a member of the committee to revise the promotion policy of officers of all ranks including promotions upto General's rank. It is extremely important that only the best and most deserving few reach the top. Defence forces have a very steep pyramidical structure of ranks. More than 50 per cent of the officers get left out at various ranks after Lieutenant Colonel due to this pyramid. Of the strength of approximately 1,200 officers who are commissioned into the army each year, fewer than 15-20 make it to the Lieutenant General rank. The army's system of selection to the higher ranks has been very fair by and large. The government of the day normally refrains from interfering in the selection process. However, the system came under severe criticism from many quarters during the early 21st century. Instances of favouritism and nepotism were alleged while selecting officers for the rank of General. The Indian Army had not fought a major war after 1971 except the Kargil war of 1999. Counter-insurgency and other forms of Low Intensity Operations dominated the period 1971 onwards. These operations are mostly Infantry-dominated. The larger share of honours and awards went mostly to the Infantry. The sheer size and span of involvement in the operations also resulted in the Infantry taking a larger number of professional courses which carried more weightage in the selection system. A need was therefore felt to review the policy, which was framed as Quantification System of Selection a few years ago.

The committee was chaired by Lt Gen. K.R. Rao, Director General of Artillery. We were required to make a presentation to the Raksha Mantri in two months' time. Views regarding the environment, of various stake holders, and other experts were taken into consideration. Our overall aim was to maintain the primacy of the confidential reports while reducing the weightage of the awards other than gallantry awards and the career courses done through nominations. In an environment where the trend of performance appraisals being inflated was beginning to adversely impact the selection system, it was a challenging task to formulate a policy of such nature. It was also a challenge to make a fair comparison and draw up a single merit among the officers of many arms whose nature of duties were varied in nature. It was Raksha Mantri Mr A.K. Antony who insisted that the subjectiveness in the selection system be reduced far more than the Committee had proposed. Promotions in the army the world over have been a cause for much debate in recent times.

It is said that Assessments drive the behaviour of the officers. If there is a need for officers to be physically fit, intellectually sharp, need to communicate well, or even the need to have battle performances for promotions, it is bound to drive their behaviour accordingly. I brought out the need for the psychological and emotional condition of the officers to be tested before assuming the command of a unit. However, other members felt that it was too early in our army to enforce such a test. Nevertheless, the Committee submitted the report within two months and it was approved by the Raksha Mantri. While conducting the selection boards for promotion to the rank of Colonels, I came across many officers who were ground soldiers and had many operational exposures but were not making it in the quantified system of merit due to their lack of certain qualifications. As Secretary of all such selection boards, I made special efforts to impress upon the other board members the importance of having motivated officers with integrity and character and operational experienced soldiers as unit commanders in the field. There were many branches of the army such as the Ordnance Corps, Electrical and Mechanical Engineers where the approval rates were barely 30 per cent of the candidates while in some minor branches such as Air Defence, Judge Advocate Branch, Education Branch and Intelligence Corps, a single candidate for single vacancy till a General's rank became the norm. I observed that the percentage of officers putting up complaints against their Confidential Reports and against negative results of the selection boards had increased many fold in recent years. The trend was not a healthy one for the army.

In June 2011, I was cleared for the rank of Major General for Command and Staff. Once again the local media in Manipur gave wide coverage to announce that at last Manipur had a General in the Indian Army. The editorial of the *Huiyen Lanpao*, a local newspaper, on June 2 wrote '...Crossing the Rubicon—a milestone...'. *The Times of India*, Guwahati and Kolkata editions dated 2 June carried the headline news as 'K Himalay becomes first Manipuri Maj Gen in

Army - A Rare Honour'. My family members were extremely happy. Making it to a General in the Army has been a journey of hard work, sincerity, commitment, and dedication to the army values. I consider this achievement is due to the blessings of the Almighty, my parents and love from my family. All the elders in the family, and my teachers have in some way influenced me in my formative years. I owe it to the troops whom I had the privilege to command during operational environments and to my superiors and colleagues who held my hand in trying times.

A General Officer in the army is a sacred responsibility and most professionals look forward to being one. American historian H.W. Brands wrote 'If you put on the military uniform, you are a prima facie hero. Generals are the epitome of that.' This succinctly explains the expectations from a General Officer. In today's environment troops and subordinate officers look up to a General Officer with awe, only if one is worth the rank one wears. In recent times, there has been criticism of the General Officers in general from some quarters over their conduct. My aim here is not to analyse the leadership of the army today at General's rank but to reflect on the challenges of being one, particularly if one is commanding large formations. The army formations today are highly mechanized and in a technology-intensive environment. An Infantry Division in operational area is expected to deal with multiple facets of a conflict situation. The expectations of the officers and men in the army have increased manifold. Man management at unit levels is also quite a challenging task due to changing socio-economic realities in the society.

Romancing the LoC!

I was lucky to have been given the opportunity to command the elite Ace of Spades Infantry Division in Poonch, Rajouri sector, where I had served in various capacities in the past. The Division is deployed in the sector in the Ceasefire Line/Line of Control (LoC) since its re-raising in 1950. This is one of the most sought after formations among the newly promoted Major Generals of the General Cadre Officers due to the challenging nature of operational environment. The Anti-Infiltration Obstacle System (AIOS) in the proximity of and all along the Line of Control had been erected by our force during 2003-05. The infiltration of militants from across the LoC had been reduced significantly in the sector from 2005 onwards. I took over the command of the formation from Gen. Om Prakash in August 2011. It was a dream come true since it was in this formation that I was commissioned in 1978 and joined my unit in Poonch. Later, I was a Company Commander in Nowshera Sector in the same division during 1990-92 at the height of infiltration by jehadi terrorists in J&K. My old friend Col Alok Joshi and Col Vipul, son-in-law of Brig. Gupta of my unit, were in my staff and they were great assets. I am proud that both the officers are Division Commanders today.

The formation was one of the largest divisions in terms of the number of units and troops under command. The defence of townships like Nowshera, Mendhar, Rajouri, and Poonch against any Pakistani attack was inherent in my task. During the period, counter-infiltration operations was one of the major challenges. We had to ensure that no infiltration of Pakistani jehadi terrorists took place in the area of responsibility. My routine comprised of frequent visits to the forward posts on LoC and units to resolve any operational related issues and also to motivate the soldiers. Frequent ceasefire violations by the Pakistani army was another aspect which needed my attention. My orders were clear. We had to ensure the Pakistani army had to bear unbearable costs for any mischief inflicted on our side. And we ensured that. Almost all the infiltration bids on the Line of Control were defeated and every militant who attempted it was killed. To my surprise I found that the militants were equipped with very sophisticated weapons and equipment. The night vision devices (NVD) used by the jehadis were definitely better than ours. They had GPS systems in place. Each militant carried most modern and state-of-the-art arms and equipment. In one case, we found a packet of narcotics on a dead body.

During a visit to Kirni village in Poonch, located on the LoC itself, I learnt that villagers who had been vacated from the village during the Kirni operations in 1992 were still in the relief camp established for the displaced population near Poonch town even after 20 years. I was serving in the same division when the Kirni operation took place to evict Pakistani soldiers from the village. The entire population of the village was living in the camp established to accommodate them back in 1992. I decided that they must return to their homes in the village. With some assistance from the Deputy Commissioner of the District and our division engineer's efforts, we were able to ensure bare essential water supply and electricity to the village within a month of my decision. My superiors were sceptical of my proposal in the beginning. But they agreed finally. The unit commanders and the Brigade Commander Brig. (later Lt Gen.) V. Sreenivas did a great job of it. The entire population of over 1,000 strong was shifted into their own village. I was also told that some of the displaced families preferred to stay in the camp itself for their own needs. Later, the Sarpanch of the village presented me with a poetry written by him on this humanitarian action of the army.

There were many shortcomings in our arms, ammunitions and equipment. Our only consolation was that the Pakistani army was worse affected by the shortfall of such items. There being no runners up prize in war, it was a felt need that we must make up the deficiencies to meet the minimum necessity for a short war if thrust on us. The matter could only be resolved at the Army Headquarters and the Ministry levels. However, it was our duty to bring these to the notice of the staff and commanders concerned. On at least one occasion I had spoken and written to the Corps Commander, Army Commander, DGMO (Director General of Military Operations) and the DG Infantry, that it was an absolute need of the hour to provide our troops with the quality level of equipment that the Central Armed Police Forces (BSF) had in the same sector under my command. It was unthinkable that the police forces designed to handle policing duties should be armed with better weapons and equipment than the army. But it was true. The reason was very simple. The government had laid down very strict acquisition process for the armed forces while it was extremely simple for the armed police forces. Hence, the decision-making was a lot easier in the Armed Police Forces. I also suspect that our establishment continues to look at the army as a legacy of the British till now. The indifference or pure lack of understanding of the army by those who call the shots is palpable in the corridors of power due to which Gen. V.K. Singh as the COAS had to write to the then Prime Minister about the 'hollowness in defence preparedness' during 2012.

I am reminded of a very interesting episode in 1992, when a Pakistani Subedar wanted to meet me at a forward post on the LoC. It is permissible under the rules to exchange courtesies during such a meeting. During the meeting, the Pakistani Subedar tried to give us a telephone set which I obviously refused. I asked him as to why the telephone. He told me that he was the commander of the post which was located about 50 yards from our forward most post. And that he being a heart patient, found it difficult to sleep at night due to the loud telephone conversations in our post at night! Hence, he wanted to present us with a better telephone set so that he could sleep in peace! It was true that we used very archaic equipment at that time. The privacy, deprivation and loneliness experienced by the soldiers in such challenging areas is difficult to narrate. Expectations and the needs of the troops have also increased multifold today. Many troops requested me to allow use of mobile phones to speak to their families. My commanders and staff were against allowing the use of mobiles in the sensitive areas due to concerns of security.

A few years later in 2015, the late Shri Manohar Parrikar, the then Raksha Mantri visited the Corps where I was the Corps Commander in J&K. During the interaction with the Hon'ble Minister, he asked me in the presence of the COAS if he could do anything for the troops. I replied that everything was fine and that the army is taking good care of the troops. On his insistence, I requested him to fast-track provision of better boots for the troops since the troops were buying their own boots and other items from the market due to extremely poor quality of the ordnance supplied boots and other items. I also told him that 80 per cent of the troops purchase their own private boots for operational use. I further added that the state of affairs has been going on since the last four-five decades and that only the attention of the RM may bring about some changes. The COAS who was present asked me if I had brought all this to the knowledge of Army Headquarters. I replied that the issue was known to all in the army and that I had spoken and officially projected the same to the MGO (Master-General of Ordnance), the DG Infantry and the VCOAS.

The COAS asked my subordinate, the Division Commander who was present at the helipad, to gather the troops nearby to physically ascertain if they indeed were wearing private boots. There were about 30 soldiers nearby. But for one or two, every jawan who was present there conveyed to the COAS that they were wearing private boots. The COAS was presumably satisfied with my remark made to the RM. The COAS then directed his staff to put up to him all the correspondence between my headquarters and the Army headquarters. I was happy that the issue would get due attention of the authorities. I hear that a number of steps have been taken or are in the pipeline to ameliorate such gaps in the army. I only wish that the future generations of senior commanders lay greater priority to the operational needs of the soldiers. The organisational structures of the formations were too cumbersome. I wish to see more agile, lethal and effective formations in future.

The period between 2005 till 2012 had been mostly of relative peace and calm in the state of Jammu and Kashmir. Fresh recruitment in the area south of Pir Panjal was negligible even though there were some recruitment activities in the valley. As per a Home Ministry report, the number of attacks on the Security Forces during 2005 was 806 as compared to over 3,000 incidences each year during the pre-2000 period. The level of terrorist activity in the division area of my responsibility was confined to the infiltration attempts from across the LoC. During one of my visits to a forward location, a group of militants whose infiltration attempt had failed waited for a lucrative target and fired at my helicopter in panic from a distance. The militants melted away to their side of the LoC immediately after firing the rocket. I was lucky. In yet another incident, a few of our forward posts were engaged heavily by the Pakistani artillery guns and one jawan suffered serious injury in a post called Nangi Tekri in Poonch District. I decided to spend some time with the troops in the same post where the jawan was injured. The post was under the observation of the enemy. After having tea in the post, as we including the Brigade Commander Brig. Yadav trekked down for a few minutes towards the roadhead, a Pakistani shell landed on the same spot where I had just rested for an hour there. No injuries. I was lucky yet again!

The hinterland upto the Pir Panjal Range was being looked after by the Rashtriya Rifles Sectors. Even there the situation was much better than a decade ago. Most of our counter-terrorist strategy was focused on counter-infiltration, de-radicalisation, winning the hearts and minds of the public at large, area domination to deny any terrorist the temptation to operate in the area. With the help of my officers in the Ace of Spades Division, I was able to compile a book titled *Romancing the LoC.* The book was released by the then COAS Gen. V.K. Singh in Delhi on 30 May 2012 when he hung his uniform after 42 years of spectacular service to the nation. The book highlights the life and times of soldiers deployed on the LoC. The book also narrates the story of a soldier serving in the LoC in repeated postings like me and falls in love with the place, people and their legacy.

There were a few occasions when communal tension spread in the area. I had gone on a few days leave for the marriage of my elder daughter during November 2011. I had to rush back to Rajouri on the advice of my superior without attending the marriage feast to quell the tension which was being built up in the town. My friendship with some Muslim and Hindu hardliners in the area came to use. Leaders of both the parties including MLAs readily agreed to my invitation. With some coercion and carrot and stick policy, we were able to calm the situation. The situation had erupted in the Rajouri area as a consequence of some inter- community clashes in Bhaderwah, another town in the Pir Panjal range.

Field Marshal W. Slim had once remarked that the most satisfying commands in the army were the platoon, a Battalion and Division. I had commanded a platoon in Poonch area in 1978-79. My Battalion command was in the Siachen Glacier, and the Division command was in the LoC. The finest commands as enumerated by FM Slim were soon coming to an end. The joy, the satisfaction of being with the troops in operational area seemed to be passing by too soon.

We ensured that the military area was the greenest spot in the entire township. During one of my unscheduled visits to the officers' mess I noticed a frame of painting of the *Badshah Nama*, the famous paintings presented to the queen of England by the Nawab of Awadh. Later I found out that some more frames of the paintings were lying scattered in the rooms of the mess. The copies of the paintings were possibly presented to the mess by one of the Gorkha Battalions a few years back. The importance and the costs of the copy painting were perhaps not known to the officers. On inquiry from some art collectors I learnt that these copies would still fetch a few crores of rupees. On learning this, the paintings were securely put up in the mess dining hall. I do believe that there are innumerable works of art and silver pieces of historical value in the possession of various army units and formations.

I was accompanied by my family during a visit to the historical fort of Chingus near Nowshera. My daughters noticed a large number of mules lined up and ditches were being dug nearby. On being asked about it, I told them that the mules would be buried in the ditches soon. All of a sudden, my daughters asked me to divert the car back to the site where the mules were. After we reached the place, they asked me to tell the officials concerned that the mules should not be killed and buried. Regulations of the army provides for casting off of old and unfit-for-duties animals, disposed by euthanasia. Much against the rules I did instruct the officials concerned not to carry out the procedure. A case was taken up for alternative method of disposal without putting the animals to sleep. However, it only saw the light of the day much later sometime in 2016-17.

The joy of flying to high altitude posts in the thick of winter snow and being served with the hot 'Langar Tea' by the troops are among the experiences which will remain etched in my memory. The innocent yet soldierly looks of the troops who continued to do their duties in spite of many hardships they face on a daily basis gave me satisfaction that our country is safe. The frequent drive from Jammu to my Headquarters at Rajouri along the many rivers and nullahs in the valley will continue to be one of the most memorable road trips in my life. My family accompanied me on a few of these drives too. Having tea in the lap of the Kalidhar Range during halts were a delightful experience. Sighting of a leopard which often squatted opposite my residence on a high ground, normally after my dinner, was simply unforgettable. Unfortunately, the same leopard got caught in a concertina wire after a month of my sighting her and died due to bleeding. I got a cement replica of the leopard made at the same spot where I first spotted her. The impressive panoramic view of the majestic snow covered Pir Panjal Ranges from my verandah was a joy forever.

Prior to our departure from Rajouri, Mangala and I went to pay our respects to 'Palma Mata', a lady who is considered very spiritual with supernatural powers. At the prayer she told me that I would return to the area soon. I wondered what would this mean except coming back as a Corps Commander! I felt that my chances of picking up the next rank were not very bright. My batch comprised of many outstanding Generals who I believe were more competent than me. Even if I were to pick up the next rank I knew that the experiences as a General Officer Commanding of a Division would always remain unique and not comparable to any other command.

Short tenures of the commanders have a huge impact on the functioning and efficiency of the units and formations. At the same time, short tenures cannot be avoided if the system has to meet the aspirations of many thousand officers who are looking forward and striving to command formations and units. To maintain balance between the organisational requirements and the individual aspirations is a challenging task. The ideal tenure for command of Brigades, Divisions or Corps should be more than 24 months in each of these formations. To meet this requirement there is a need to further reduce the approval rates in the promotion boards from approximately 50 per cent to 35 per cent or so. This will further increase the dissatisfaction levels in the Army. Also ever-increasing number of non-empanelled officers (those who do not get promoted to the 'select rank' of Colonel, mainly due to limited vacancies) from roughly 5,000 to more than 7,000 within the system remains a cause for concern. Many measures have been considered by the Army Headquarters in the past and implemented to address the issue. All these efforts appear to be more cosmetic in nature and there is certainly a need to remodel the entire Human Resources issue of the army. Sooner the better.

Tryst with the Western Deserts

I joined the elite Desert Strike Corps as the Chief of Staff (COS) in early 2013. The major components of the Strike Corps are the tank units. I had never been inside a tank nor operated with tanks in the past when I joined the corps. I trained myself in the basics of a tank. Lt Gen. Amit Sharma who was the Corps Commander and I had served together in the Gangtok Division. It was a pleasure to be colleagues once again. We conducted many training exercises and discussions on the operational front. I learnt many new things in the tank and desert environment where our role was during war. Gen. Bikram Singh, the then COAS, attended one such training event in which I was performing the role of the Corps Commander of the Strike Corps in battle. I have had no interaction earlier with Gen. Bikram Singh in my entire army service. It was important that I performed well in the event. At the end of one full day of discussions and deliberations it was a great pleasure to be complimented by the COAS on my performance in the event. My tenure as the Chief of Staff of the Desert Strike Corps was an eye-opener for me. I was fully convinced that the Corps would deliver in any event of war on the Western Front.

My promotion board to the rank of Lieutenant General was only two months away. The unforgettable complimentary remark in my favour by the COAS in the presence of my bosses and subordinates made me feel that my next rank could be after all, a possibility. There was speculation in the air that the perceived sour relationship between the former COAS Gen. V.K. Singh and Gen. Bikram Singh could affect my chances for the next rank. Such rumours and speculations do take place in any organisation. But the wisdom and maturity of the persons concerned seldom allows such biases to come in the way. The Special Selection Board (SSB) and the Appointments Committee of the Cabinet cleared me for the rank of Lieutenant General in 2014 in the Command Stream along with 11 others. Those who were cleared for the Command were my friends and coursemates, one better than the other. Accordingly, I was ordered to relieve Lt Gen. D.S. Hooda in the Nagrota Corps in J&K and Hooda was sidestepped as the Army Commander Northern Command.

The First Lt Gen. of the Army from North East India

The stars had connived to destine me to become the first North Easterner to be made a Lieutenant General in the Indian Army. The North East has produced two Air Force three stars from Assam, including Air Marshal Anjan Gogoi, elder brother of former Chief Justice of India and a Sainik School Goalpara alumni. It was a flattering experience to see *The Indian Express*, Delhi edition, put my promotion story on the front page. Many complimentary messages poured in from various dignitaries, including one Governor and two chief ministers of some states in the North East. How I wished that my parents were alive to hear the news. My wife and daughters have borne the brunt of separation during my army career. My wife has brought up the girls almost single-handedly in my long absences due to operational reasons. My brothers, sister and other relatives were celebrating my promotion too.

The period between 2014-15 at Nagrota in J&K was relatively peaceful for the Army. Militancy in the south of Pir Panjal was already on the ebb. I was fortunate to have served in the area multiple times in the past. I settled down quickly to the new place. The official residence of the Corps Commander is named 'Kopje', a beautiful house on the banks of Tawi River. The Tawi River is at least 700 metres wide in front of the Kopje. The house was named 'Kopje' by Lt Gen. J.F.R. Jacob, PVSM, who was the first Corps Commander of the Nagrota Corps and later became a Governor of a state. In Tunisia, Africa, 'Kopje' means 'a small hill in a generally flat area'. Gen. Jacob perhaps named the house in memory of his war campaigns in Tunisia during World War II. From the lawns of the house the Tawi riverbank provided a most picturesque view. Most members of my immediate family visited us during my tenure there. As the Corps Commander I was also one of the Ex-Officio Security Advisors to the Government of J&K. There was regular interaction with His Excellency Shri N.N. Vohra and Shri Omar Abdullah, the then Governor and Chief Minister of the state, respectively. I was grateful for their valuable inputs to enable me to see things in the state from their viewpoints.

The flood in the state during September 2014 was one of the worst in recent years. 'Operation Megh Rahat' was launched to save lives and property in the entire region. Thousands of homes were destroyed. Thousands of residents were displaced. Our troops carried out flood relief in the most trying circumstances. I recall one night it was reported that about 30,000 people were walking from the valley along the road towards the Banihal Pass and beyond to Ramban. It was a mammoth task to manage the huge crowd of hungry people who migrated from the valley en masse. There were roadblocks due to mudslides at many places. Immediate action was needed to feed the crowd, provide medical attention and to evacuate them to safer places with some assistance from the civil authorities. Next morning all available helicopters were deployed for their rescue and to drop food items for the stranded crowds at various locations along the highway. Medical teams and rescue parties were pushed into action early morning. It was easier for the army to respond since we were physically deployed all along the Srinagar Highway. And we had the resources to help the administration. Later it turned out that the count of the crowd inching towards Ramban was almost 50,000 in strength. Meanwhile 'Kopje' was also affected by the flooding Tawi River.

In the anti-terrorist operations there were many encounters where many jehadi militants were eliminated with ease by the forces. I often shared my thoughts as to the reason why the Muslim majority areas in the Pir Panjal Ranges and the Chenab Valley areas remained peaceful while Srinagar valley was always in the news for violence. Kishtwar, Gulabgarh, Gund, Doda and other such towns were in the lap of the Pir Panjal, neighbouring Srinagar valley. These areas had sizeable Kashmiri population even though the Gujjars, Bakarwals and Pahari Muslims were the majority. Though these areas were in proximity to the valley, they remained physically isolated due to snow in the Pir Panjal Ranges except for the passes which were well guarded by the army. The resources required for militant activities were not available in the foothills and in the south of Pir Panjal Ranges. During the period 1990-2004, before the LoC wire fence came into existence, the resources to assist the militants were flowing in abundance from across the LoC. The LoC fence which came up in the Corps Zone made this flow of resources much more difficult for the militants. The radical groups were not able to radicalise the migrating tribes of Gujjars, Paharis and Bakarwals who comprise the majority in the foothills of Pir Panjal Range. The militants attempted multiple times to use tunnels in the Akhnoor area to infiltrate to our side. But each time their nefarious designs were defeated. The trans-Pir Panjal road along the erstwhile Mughal Road which connects Srinagar with Jammu through Rajouri and Shopian was also well guarded against militant activities.

Radical Islam was yet to make full impact along the foothills south of Pir Panjal even though cases of radicalisation were reported in areas like Mendhar, Poonch, Doda, etc. The mixed population of many tribes in the area did favour the Security Forces. The narrative in the valley was beginning to change during 2013-14 for the worse with the idea of ISIS and Caliphate raising its head. During this period the use of the internet as a weapon of the terrorists was clearly visible. Social media became a tool for terrorists' activities, particularly to recruit young people and to propagate their ideologies. Social media propaganda became more and more effective among the militants as exemplified by the rise of young people like Burhan Wani during 2014. Thousands of militancy-related pictures and videos consumed the Kashmiri youth and propelled them to join the *jehad* as they were highly inspired by the propaganda machine through the internet. A new phase of *jehad* in the virtual world thus emerged in the state sometime during the period. Social media was also used by militants to discredit the state and glamorise their narrative of gun culture among the local youth. While we in the Security Forces were counting the number of militants neutralised by us, the likes of one dead Burhan Wani mobilised hundreds of subverted minds.

In one instance of the Pakistani army's violation of the Ceasefire, I ordered a far tougher response designed to deter the enemy. I also warned that 'unexpected damage' to their forces will be inflicted if they continued with such ceasefire violations. The statement was prominently covered in Pakistan's *Dawn* newspaper on 22 July 2015. It read:

> Indian Army's Lt General K.H. Singh warned, 'There are certain elements in Pakistan who want to create trouble on the LoC... We have to give unexpected damage to them so that they don't repeat it in future'.

The statement was trolled on social media by hundreds of Pakistani youth in the next few days. Ceasefire violations on the LoC were nothing new. We were prepared to give very strong responses at that time. The year 2015 saw an overall increase in the intensity and the scale of heightening tensions in the state and the LoC. The situation worsened further in the LoC and the Indian Army was forced to carry out Surgical Strikes across the LoC in September 2016 nearly a year after I left the Corps.

Counter-infiltration operations in the Srinagar-based Chinar Corps in the north whose operational role comprised the whole of the Valley and the mountains surrounding the valley, were more challenging in view of the tougher terrain there. While infiltration of militants in the area south of Pir Panjal was almost reduced to zero level, the same could not be said in the area north. The policies of the Central Government towards the Kashmir problem have also been wavering from time to time depending on the personalities involved in the process. In fact, there never was any Kashmir policy in the true sense that I knew of. It was the Security Forces at the tactical level trying to stop infiltration, to neutralise the terrorists, patrol the cities and towns, and win the hearts and minds through 'Op Sadbhavana', a small measure to engage the young in schools and colleges. Serious engagement at the political, diplomatic, and at the socio-economic front was missing for many years. No attempts were made towards de-radicalisation of the youth.

In other words we seemed to focus more on the tactical success of killing terrorists while the more important issues were left unattended by those who ought to have done so. Sun Tzu wrote on the art of war, 'Tactics without strategy are the noise before defeat'. Kashmiris as individuals were completely different from the group behaviour that was displayed in the streets. The Kashmiris are friendly, God fearing, soft spoken and endowed with all the good qualities of the Sufi culture. However, public discourse and violent activities of throwing stones at the army and police forces presents a different group behaviour pattern mostly influenced by the radicalised ideology.

There have been frequent comments in the print and social media about the need to revisit the Counter-Insurgency (COIN) Operations Strategy of the Indian Army. They claim that the strategy adopted by the army is mostly archaic and does not suit the modern realities. While I partially agree with the comments, it needs to be clarified that these strategies or tactical training in our training establishments are only given as guidelines for the units and formations. Formations adopt their own Standing Operating Procedures (SOP) keeping the local conditions and complexities in mind. These SOPs at the tactical levels have been found to be working in most cases. In a conflict situation the commander or the leader on the spot is permitted to exercise his or her discretion to handle the situation keeping in mind the basic tenets like minimum use of force. The initiative on the spot is the key. For that the junior leaders are exposed to varied conditions under which they are likely to operate while undergoing training in the schools of instructions. Ultimately, actionable intelligence is the key to such operations. For this, the population, the state police and the machinery have to be largely on the same side.

There are opinions and debates over the issue of relevance of war in any form or not. These learned people forget that wars are but an extension of the polity of the nation. Wherever and whenever national interest of the country is adversely threatened there will be conflicts of varied nature. As and when conflict resolution mechanisms fail to achieve the national objective, it is likely that war, conventional war, limited conflict or other forms of hybrid warfare are bound to occur between nations. The armed forces are designed to prevent war with other nations and to win if war is thrust on them. I agree that the modalities of achieving national objectives can be varied and debated. But to assume that other countries will not wage war, just as Nehru once believed, will be suicidal. Armed forces need to evolve and transform to meet the future multi-dimensional threats. To this end I feel that we are lagging far behind. In this context, the Role of the Military will need a serious relook. The military together with other security establishments needs to go to the drawing board again to consider the emerging challenges around the world afresh. The decision loop in the army appears to be concentrated in the South Block, while giving a semblance of consultations with the field commanders in mundane matters. A top down organisational transformation to empower the lower formations with clear cut National Security Objectives is imperative. At the same time, the need to integrate the resources of the state to counter the emerging security challenges should be addressed.

Pakistani Army and its Obsession with Kashmir

My impressions of the Pakistan Army during the period were that they were ill-equipped and ill-prepared to fight a full-scale war or even hybrid war with India. The radical groups within the Pakistani army have grown in strength. Standard of military training appears far below the standard required for a modern army. The army appears to be pulling in two different directions, of Islamic ideas on one side and the military profession and values on the other. Their obsession with Kashmir could be their Waterloo in the hands of the Indian Military. Kashmir, according to them, is their existential question and the cornerstone of their foreign policy. We need to accept the fact that the Pakistan Army will never allow India to remain in peace. The morale of their troops and unit cohesiveness appears to be low too.

I recall that at least two soldiers of the Pakistan Army had intentionally crossed over to our side to escape from the physical and psychological rigours in their units which were deployed on the Line of Control. One soldier who deserted towards one of our posts in Poonch confided to me that he was troubled by the others in the unit, being a Baluchi. He was repatriated to Pakistan through the Chakan-Da-Bagh crossing in the Poonch sector. However, one should not forget the fact that a regular individual Pakistani soldier is a good fighter.

With the Islamisation of the army, the will to fight India may be stronger due to the radicalised nature of the Pakistan Army doctrine at present. The Pakistan Army's dependence on Chinese and American assistance also needs to be factored in. China is likely to help Pakistan in keeping the fire burning in J&K to bleed India for its own interests. India's diplomatic proximity to the US and Israel in recent years could also bring the Pakistan-Taliban relationship to a higher level. Hence, the dynamics of security in J&K is dependent on the dynamics at the international level and regional powerplay.

I heard stories about various options being deliberated to resolve the Kashmir issue—some mention about a Chenab formula, LoC being made irrelevant, joint monitoring mechanisms and so on from time to time. Having been a field commander from the lowest level to the highest during a major period of my army service, I am of the view that none of these are likely to work on ground in the long term as long as the military controls the polity in Pakistan. These proposals appear to be more of diplomatic jugglery, full of sound and fury, to initiate and begin talks. The way forward could be increasing the pressure on the Pakistan Army, hurting their jehadi economy further to ensure that their people reject the rule of the military by proxy. All these efforts need to be backed by solid military capability to deter Pakistan from initiating any form of sub-conventional war against us. At the same time, effort to strengthen the rule of law, eliminate the intruders and bring people of the Kashmir region closer to the national mainstream, should be made. Time is on our side too. We have a big stake in ensuring that efforts to solve the J&K issue by resorting to undue concessions to Pakistan do not bring about more divisive forces to the fore within our country. The idea of Islamic State of Khorasan (ISK) lives in the minds of the radicalised souls in Kashmir more than anywhere else in the country. US Marine Corps General Kenneth McKenzie had also told the Senate Armed Services Committee in

December 2018 that Pakistan continues to use the Afghan Taliban as a 'Hedge' against India.

I am also of the view that migration of some highly radicalised elements from Afghanistan-Pakistan region towards Kashmir is a real possibility once a peace deal with the US in Afghanistan which came in February 2020 is enforced. The ISIS with a formidable number of approximately 6,000 fighters could direct some of their cadre to the valley in the foreseeable future at an opportune time. Even the Pakistani army would be able to shift over 50,000 troops from its western borders to the east. India is far better prepared today in terms of troops, police and polity than it was in the aftermath of the Soviet withdrawal from Afghanistan 30 years ago. Peace in Afghanistan even after the US and Taliban accord appears remote as things stand today. The reported interest of Iran in the Shias of Kargil in Ladakh region is also another factor to be taken into consideration to avoid a Sunni-Shia battleground in the region.

As I pen this today, 6 August 2019 the news of abrogation of Article 370 in Jammu and Kashmir reverberated across the country! A hard decision of this nature was what was required in the state. The caveat being that the country needs to be prepared for all the possible consequences in the Indo-Pakistan relationship and the world opinion. Pakistan views these developments as the end of the Simla Agreement and potential conflict in the LoC under a nuclear overhang, and also sees the development as an opportunity for intensified terrorism in the state. It is difficult to predict the short-term future trajectory of the jehadi militancy. The Indian state will have to be prepared to grapple with the fallout of the abrogation of Article 370 by way of proactive Perception Management, psychological ops, political and diplomatic initiatives, and above all by ensuring that the pillars of the state machinery remain intact. The Armed Forces and other security establishments like the CRPF, J&K Police and the intelligence agencies will need to remain ever vigilant to tackle the emerging security challenges in Jammu and Kashmir. 'People First Policy' should go hand in hand with 'Nation First Policy' in the future.

Goodbye to Panchal-Desh

All good things come to an end sometime or later. The tenure of Corps Commanders is mostly limited to one year. A Lieutenant General with two years residual service after Command of a Corps is assigned as Army Commander, subject to there being a vacancy, and those who don't have the same are sent on staff assignments or to schools of instruction. Therefore, I was not to be assigned the responsibility of being an Army Commander. However, any ex-Corps Commander who was not assigned the higher responsibility purely due to age factor is granted the pay of the higher grade. I have had a memorable tenure at Nagrota as Corps Commander. The eternal sight of the abode and shrine of Vaishno Mata in the horizon as my helicopter took to flight in the morning almost every alternate day was a most spiritual sight. The dark clouds in the narrow valleys as I flew to various far flung locations were pure delight of a lifetime. Having been a mountaineer for years, I could not have had a more enjoyable tenure. Meeting and interacting with thousands of officers and men under my command gave me more faith in the inner strength and capability of our officers to win in any situation. Only the Generals have to live up to it.

When the time came for me to bid farewell to the White Knight Corps, I made a request to the COAS and the MS (Military Secretary) to consider me for Kolkata or Udhampur where no other Lieutenant General rank officer of the General Cadre was willing to go. I believe that Gen. Dalbir Singh Suhag insisted that I go to Mhow. He must have had very special reasons to send me there. In sum, there have been many successes, while some projects remained incomplete. Much as I pushed for better equipment and clothing for the troops, things moved slowly and improved to some extent a few years after I left the Corps. Governor N.N. Vohra wrote a very touching letter to me when I left the Corps. He had been a pillar of wisdom for me during my assignment in the state. Col N. Romeo Singh, my Colonel Military Secretary, and Capt. Praveen Chaudhary, my ADC, had been of immense help to me.

The Infantry, Mhow

On completion of 16 months in Nagrota, I moved to Mhow. My wife joined me there. The 100-years-old Infantry House from where I was to hang my uniform in February 2017 was located opposite the Infantry War Memorial on the Mall Road in the Cantt.

It was a wonderful opportunity to interact with the officers of the Infantry and other ranks. Various infantry specialised courses are conducted in the institution. Mhow is like a Mecca to the Indian Army's infantry, the 'Queen of Battle'. I recall a talk by a senior General of the Armoured Corps way back in 1989 during the staff course at Wellington who made a statement that those officers who were not acquainted with Mechanised warfare will be a liability in the army or words to that effect. Meaning thereby, that relevance of the Infantry is slowly coming to an end. As we have seen in the recent wars in Afghanistan, Iraq or even in Kargil, Infantry will continue to be the decisive arm, no matter what level of technology is fielded in future wars. I wish to see more tech-savvy Infantry in the near future. Finally the boundaries of nations and countries will be decided by the boots on ground. Talks on the irrelevance of national boundaries in the future are a welcome change for debates but not pragmatic at all.

My priority was to impart highest standards of individual skills in tactics, use of weapon systems, to operate as part of small teams and overall grooming of the Infantry officers. The individual initiative, flexible manner of operations to fulfil the assigned task, were also given maximum weightage. The officers posted in the Infantry School were among the finest and the most professionally sound in the army. Some additional courses were also introduced during the period to enhance the capability of the junior officers in the units. Greater emphasis on use of technology during Commando training was attempted. New short courses for officers and men were introduced too. I had planned to complete at least 20 per cent of the Infantry Museum Project. The building was complete but the making of the museum was a long way ahead. The museum staff worked tirelessly. Col Biswas who was in charge of the museum ensured that the target was achieved when we were able to open the completed ground floor for public viewing on my last day in uniform.

The Question of Civil-Military Relations: A Matter of Honour

An incident involving some young officers (YO) and the civil police in Indore city deeply concerned me. Some YOs had gone on 'out pass' from the premises at Mhow to the city. The police, led by one Additional Superintendent of Police, entered the bar where these YOs were having a drink. The police asked the bar owner to close the bar after its timing of 10 p.m. Some arguments ensued and a physical scuffle took place where four YOs were severely injured after being thrashed by the police contingent. Rest of the YOs returned to their barracks only to return with more number of YOs to Indore and were looking for the policemen who beat up the officers earlier. These YOs entered the police station and chased the policemen on duty there. The matter became a national news.

I rushed back to Mhow from Delhi and tried to resolve the issue. I met the Chief Minister, the Chief Secretary and the DG police of Madhya Pradesh. The Chief Minister was of the view that the matter should be amicably settled between the army and police. However, the police felt that such incidences have occurred a few times in the past and they were of the opinion that the army people should be taught a lesson. They insisted that we hand over the errant YOs to them for inquiry. They were politely told that these officers cannot be handed over to them. We suggested a joint inquiry to which they claimed that there was no such thing as joint inquiry. We stuck to our stand. The police lodged FIRs against the YOs with very grave non-bailable charges like attempt to murder etc. In response, the YOs also lodged an FIR which the police refused to entertain. The case came before the High Court which gave the verdict after a few months in favour of the army. The state police went to the Supreme Court later. The Hon'ble Supreme Court also upheld the verdict in favour of the army personnel and indicted the civil police for non-filing of the FIR as well as for police high-handedness. This issue was important from the point of view of the position of the army in the society, the attitude of the civil police in such circumstances. These incidents do no good to the image of the police nor to the army. The larger issue is also the concept of civil-military relations at various levels which appears to be on a downward trend even today. Many suspect that the Nehruvian fear of the Armed Forces still haunt some people to this day.

There have been many debates about the One Rank One Pension (OROP), grant of Non-Functional Upgradation (NFU) to the defence forces officers, downgradation of the defence forces in each Pay Commission and many other issues. I do feel that the responsibility lies on the shoulders of the political leadership given the fact that the civil services are most unlikely to accommodate the military in what they consider as their fiefdom. It is therefore imperative that the political establishment take some effort to understand and take action to see that the Armed Forces are given their due honour in the society. At the same time, military veterans also need to see the changing realities in the country. A strained civil-military relationship can be a disaster for the nation. I am also of the opinion that the

Armed Forces themselves have a lot to explain in their inability to command the confidence of the political leadership. The military could also be faulted for their inability to put across their logic in proper perspective in sensitive matters. We are yet to see leaders of the kind in the contemporary Indian Military leadership who are able to put their rank on the line, if need be, in the interest of the organisation.

More of Generalship

In olden times the difference between civil and military leaders was narrow. In fact soldiers and statesmen were interchangeable. For many years the rulers of the state led the armies in the field. Today, political Generals are anathema to the Indian Military, a belief system inherited from the British. Yet in the British history Cromwell, Marlborough, Wellington, and Churchill delivered the greatest military victories even though they were either Parliamentarians or in politics and were latecomers to the military. In his classic study, *The Man On Horseback: The Role of the Military in Politics*, Samuel Finer states that 'The moment the military draw this distinction between the nation and the Government in power, they begin to invent their own private notion of the National Interest'. The Indian Armed Forces have been amenable to Civilian Control of the Military. But, the issue in India is, people tend to confuse this relationship with bureaucratic control of the military rather than by the elected political masters of the government. On the other hand, countries with overbearing military have failed miserably. Resistance of the Pakistani army to let democracy take roots in that country has failed the people of Pakistan. Bangladesh went almost a similar way a decade ago. Myanmar hopes to recover from the effects of decades of military rule.

According to Socrates, a General must know good administration, make plans and be able to carry them through. He should be observant, untiring, shrewd, kind and cruel, simple and crafty, a watchman and a robber, lavish and miserly, rash and conservative. He should also know his tactics. The order in which Socrates has put the qualities of a General sounds strange. He put Administration first and Tactics last. Our army puts overriding priority on Tactics. The priority changes with time. In one of his lectures, Field Marshal Earl Wavell put physical strength and moral character, and mental qualities like knowledge of the mechanics of war, including the logistics of war as priorities. He also mentioned that it is the knowledge of the mechanics of war, not the principles of strategy that distinguishes a good leader from the bad (*Wavell, Soldiers and Soldiering: Or, Epithets of War,* 1953). The focus of training of our senior commanders as it exists today is far from the wisdom of Wavell, Socrates or any of the great minds of military affairs.

I am of the firm belief that our commanders need more lessons on the entire gamut of administration and on human resources development in the true sense. Too much of emphasis on strategy is being given in the courses of instructions even at the levels of junior Colonels. It is important to understand the strategy but not at the cost of tactics and of leadership in all forms of warfare. Strategy is the art of taking forces to war in a favourable position while tactics is about winning the battles to win the war. A renowned journalist friend of mine once told me that strategy is best left outside the field forces and we should concentrate on the warfighting! I was a bit surprised by his comment since he was one of the few journalists who backs dignity and place of the soldiers in society. I am of the opinion that very senior commanders need to be thoroughly clear on National Security Objectives and the strategy to achieve those objectives. Tactics is far more difficult to practice on ground and is a more important task of the commanders in the field. A General needs to understand and act, both at the strategy and tactical levels. Above all, Generals are responsible to ensure that region, race, religion, caste and creed are not allowed to debase the organisation under any circumstances. It is the bounden duty of Generals to preserve the culture and the ethos, to defend the Idea of India and the Values of our Nation. For that, a General needs to be a soldier with Courage, Character, Competence and Connect (the 4C's) with his troops and subordinates.

Take the Decision General!

There appears to be reluctance on the part of some decision makers today to take decisions. In a crisis situation, the worst thing is for leaders not to take a decision. Many have perfected the art of transferring the decision to the next senior, and the chain never ends. The buck must stop at the table of a General. The next worse thing, of course, is taking a wrong decision. The General too must bear the consequences of his wrong decision, if any, and not to pass it downwards or upwards. Most leaders refrain from taking decisions due to organisational constraints or due to fear of failure and insecurity as well. Some decision makers get confused between what is important and what is urgent. Some other decision makers simply do not have the guts or the knowledge and wisdom to take the decision. Worse, some refrain from owning up the responsibilities emanating from their own decisions. Visionary leadership is what is expected of the Generals in command of troops. Looking into and understanding the future, preparing for unforeseen eventualities, and having the ability to clearly chalk out the way ahead under the circumstances and do what is in the best interest of the country, is what a good General is made of.

Grooming of Middle and Senior Level Officers

The idea of grooming the senior officers in the military was highlighted by me at forums from time to time. In a first, I wrote an article on the subject in the June 2001 issue of the *Infantry* magazine. I did not find many takers for the contents of the article in the environment.

A vast majority of officer cadre at the middle levels find inadequate professional work to tap their full capacity and contribute to the organisation. Most officers hardly read or write meaningful thoughts on the military. I made serious efforts to encourage reading and writing among the officers. Hundreds of books on military subjects were personally handed over to individual officers to read and write reviews which were to be sent to me directly. Book prizes were most common prizes at officer events. Discussions and debates on military matters were frequently conducted. The aim was to have an environment where officers were forced to think and express themselves without fear of any adverse fallout. It may come as no surprise that over 40 per cent of officers claim disability pension. Keeping oneself physically fit is no longer a motivation in today's military. Too many officers doing the same thing at various levels makes their work redundant. Intellectual collapse of any military is the beginning of the collapse of any nation.

The period between 2003 and 2017 saw emergence and serious debate on what military experts claim to be 4th GW (Fourth Generation Warfare), OOTW (Operations Other Than War), hybrid, non-contact/kinetic operations such as cyber warfare and so on. There are senior politicians, civil servants and military officers who are of the opinion that fighting wars between countries is a thing of the past. While respecting their views, it is imperative to understand the geopolitical realities of our country especially with almost 50 per cent of the borders with our neighbours contested. The aim of all strategy will continue to be 'War Avoidance' as the first priority, failing which nations have to defend themselves against inimical actions in all forms of warfare. Therefore, in our context both contact and non-contact wars have to go hand-in-hand in preparing our forces. Officers will need to be a warrior against any form or types of war. The foundation courses about such types of warfare have to be laid in the institutions such as the Infantry school, War College, etc.

5
Nation First

Old Soldiers Just Fade Away: Manipur Calling

On 27 February 2017, I was given the customary farewell dinner in the Infantry Mess at Mhow. My 39 years, nine months of service in the uniform came to an end on that day. On the same day, I had the privilege of inaugurating the 30-feet-tall Bayonet (a symbol of Indian Infantry) and the Infantry Time Capsule placed underneath the bayonet. All present there resolved to meet at the same venue 10 years later! I also had the privilege of inaugurating the Infantry Museum with 20 per cent completed work as planned (opened to public view the following day) at Mhow. The Army Brass Band played the tune 'Auld Lang Syne' as I drove to the Infantry House for the last time in full ceremonial uniform.

I noticed the dim lights of the Infantry War Memorial in front of the Infantry House as Mangala and I drove past. The final and formal wreath laying at the Memorial was done earlier in the morning. In attendance were many retired Generals residing in the station. Indeed a touching Farewell to Arms. Mangala and Priyanka had organised a befitting retirement dinner in the House the day before. Mangala had made many friends at every station we were posted in. The ladies liked her charm and demeanour. We all know that someday we have to shed our uniform and bid goodbye to the army. Yet, when the day actually comes it is a very nostalgic occasion and evokes mixed feelings. The Infantry House ceremonial guard of the Sikh Light Infantry had lowered the Infantry Flag the previous evening to the bugle call tune of the 'Retreat'. We were seen off by Maj. Gen. Rawat and all officers and their ladies, and personal staff present over a cup of tea in the lawns. I knew that we would be missing them all.

As Mangala and I were driven to the airport at Indore, both of us looked back at all the historical buildings, the officers proceeding to their training areas in their cars, motorcycles and bicycles, the officers mess, the Mhow bazar, the living accommodation of the families. And quietly the routine carries on at Mhow again. Everything settles down in nature again no matter who may come or go. This is also true in the army. Many years ago a senior subaltern told me that nobody in the army is indispensable. Another guy will do an equally good job. It was true.

At the Imphal airport, my younger brother Santosh waited for us in his car just as he had been doing for many years in the past. He took us to his house to stay till a house was ready for us. I had not thought of any definite plans. Post retirement, I was preparing myself for the PhD viva at the Central University of Jammu. I had applied for positions in the Armed Forces Tribunal, and Information Commission of India too. I later learnt that none of my applications were even considered or recommended by the Army Headquarters for some unknown reason. Some local newspapers in Imphal wrote about my future prospects post retirement in some higher responsibilities in the state or elsewhere. I had no such illusions. The reality is that most times one needs to be in tandem with the powers that be and work as per their script if one has such grand ideas. The soldier in us, the belief system and the values I cherished as a soldier also keep me away from the pursuit of positions by other means. I had known many people in high places but never compromised on my self-respect.

My mother once told me, 'Do not compromise yourself for anything. Nothing comes free'. The question of being offered anything in the Government at the Centre or the state seemed remote in any case. I often wondered 'What now'! My heart told me that I had to keep the promise I made to the late Shri Abdul Kalam, ex-President of India. My promise to him way back in 2007 was that I would settle down in Manipur after retirement from the army. He never forgot my promise till a few years before his passing away. He wanted me to interact with school and college students from time to time in Manipur. He also spoke about the moral courage of the leaders of Manipur or the lack of it as he saw. He knew that I may have no big opportunities in Manipur. Having reached Manipur I started wondering about the direction of my future. I toyed with the idea of establishing a school. But it required a considerable amount of investment.

We had moved some household items to Greater Noida to make the house there liveable for visits to the National Capital Region. Our small hut in Imphal and ancestral house in Charangpat village were not in liveable condition. It took about 20 months to construct a house adjacent to my brother's house. The cost of land in Imphal area is perhaps one of the highest in the country due to the massive pressure on land in the valley. The land price rates continue to go upwards every month due to higher demand from black money sources. Well-to-do people with huge black money reserves have also reportedly bought huge chunks of land in the valley which is said to have pushed the real estate rates sky high.

Savings of my wife and our retirement benefits were hardly adequate to buy a small piece of land and construct a modest house. I estimate that it cost us 20-25 per cent more to construct a house in Manipur as compared to the nearby city of Guwahati. The cost of building materials goes up significantly due to the illegal taxation imposed by the militant groups all along the highways linking the state to the rest of the country. The might of the Indian state has not been able to contain the illegal taxation which raises the cost of living in Manipur by 25 per cent. Low quality construction materials dumped in the markets in Imphal was another problem area. It came as a surprise to me that most of the dealers do not provide cash receipts. Most of the dealers were not paying correct amount of tax to the authorities, and the latter were found to be content with the arrangement. The concept of contract system for building private houses was virtually non-existent. Hence, Mangala and I had to be physically present most of time during the construction work. We managed to create a small room to house all the military mementoes, medals, war souvenirs, and other artefacts which we had collected over a period of 40 years in the army. We completed the house and shifted there a few days before Priyanka's marriage in mid-December 2018. By this time Priyanka had joined a government hospital as a child psychologist and Poornima was in San Francisco.

I could open many photo albums and reminisce of my days in the army only after shifting to the new abode. I was apprehensive that it could be extremely difficult to adjust to the new environment. But later it turned out that all my fears were only delusions. My eldest cousin Sarat, retired Professor of Manipur University, asked me how I was going to adjust to the new life. I simply told him that I would begin a new chapter where there was ample opportunity to rediscover myself. I also told him that I would simply assume that all my glorious and adventurous life and times in the army were but one chapter of my life, and to take the same as 'just another past dream'. Every retiree experiences the retirement blues sooner or later. A life where the day-to-day struggle begins.

Mangala and I were fully conscious of the reality that all the attention and perks we enjoyed while in service were due to the position I held. These were just another facet of life. None of those were meant for us as persons. It was meant for the position and the responsibility one held. We were grateful to all those who showered me and my family with love throughout our life in the army. We made innumerable friends along the way. We also knew that many seniors and juniors and their families would remain in our hearts for a long time as many of them touched our hearts. As I write this today, we are grateful that many colleagues from my school days, NDA, NDC, regimental associates, and many other colleagues remain in touch even today.

Some of my childhood friends from the school days had expired due to some illness or the other. I had lost contact with many others. Some of my Sainik School Goalpara friends were in touch off and on. They welcomed me back to Manipur. I had a few friends from the veteran community in Manipur. Some of them had been long associated with me in my army life. Many of the veterans I knew had settled outside the state in places like NCR, Pune, Bengaluru, etc. In Manipuri society, one of the means to interact with friends and relatives is at religious/social ceremonies. Social life is mostly restricted to these ceremonies and functions. There are no army or civilian clubs worth where one can enjoy get-togethers and other entertainments. I had lost contact with many of my extended clan members. Therefore, I had to build on my restricted social circles post retirement. My younger brother Santosh made sure that Mangala and I were taken to meet old relatives regularly. We also travelled to every nook and corner of Manipur to re-familiarise ourselves with the new environment.

Post Retirement: Learning Never Ends

I received my PhD degree in National Security Studies from HH the Dalai Lama after my retirement. It was a rare opportunity to be blessed by the holy man. My thesis was titled 'Ethnic Conflict in Manipur'. I was the first recipient of the degree in the university since the department was only three years old. One of the most important findings in my research was that a large number of hill men wanted the valley people to reach out to them for a peaceful resolution. I often wondered how exactly this was to be initiated. The 'Go to Hills' slogan announced by the Biren Singh government is a good beginning. The slogan has to be converted into action to have some impact on the ground. I received some theses for evaluation from some universities as well. Post my retirement from the State Public Service Commission, the evaluation of PhD theses engaged my ample time at home. I feel proud to be associated with the Department of National Security Studies (NSS), Central University of Manipur, as an Advisor and as Visiting Faculty.

I envision that the Department can act as an academic platform for multi-faceted security issues in the region. The Department can also act as a think tank for security related issues in North East India. India's Act East Policy will require a peaceful and conducive security environment in the North East. Manipur University is often caught in the crossfire of politics, contract works and varying ideologies of all hues. I do hope that this Department gets to the position that I wish for the good of the people of the region. Today, the discourse on security related issues is dominated by the media, CSOs (Civil Society Organisations) and some NGOs. The NSS Department can play a significant role in going into the depths of the issues involved. Short capsules to familiarise themselves with the security establishment of the country from a wider perspective is a must for the media, selected officials and the enthusiastic public. Memorandums of Understanding

(MOUs) with the defence forces, universities and the various think tanks will be imperative.

A Constitutional Responsibility: An Opportunity to Root out Corruption

During the month of April 2017, a suggestion to become the Chairman of the Manipur Public Service Commission was made by the Chief Minister of the state, Mr N. Biren Singh. I gladly accepted the offer. Many organisations and well-wishers were against my taking over as Chairman of the Commission. Their concern was that my image and the reputation built over 40 years of exceptional service in the army could be tarnished since the reputation of the Commission was at an all-time low for all the wrong reasons. I was flattered by their observation and concern for my reputation. I could calm them down by putting across the counter argument that if any organ of the state becomes soiled, it is our duty to clean it up. I therefore accepted the challenge to clean up the system and put the reputation of the Commission back on track. They saw some reason but were not fully convinced. I was also fully aware that any chance of my being engaged at any Central Government level was almost zero. What better than to serve my own state when the opportunity to contribute to the society was at hand?

I joined the Commission as the Chairman on 17 May 2017. My classmate in the Sainik School, Goalpara Mr W. Hangshing, IAS, and Mr J. Shimray, IAS, also joined the Commission as members in the same month. There had been no Chairman of the Commission after November 2016. The Civil Services Recruitment process of 2016 was under scrutiny of the High Court due to alleged irregularities in the examination process. No recruitment of any kind was carried out for a long time. No Departmental Promotion Committees (DPC) for the 42-odd departments in the state government had been held for long. Many recruitment processes were held in abeyance midway by the High Court. The court cases were far too many to handle. More than 8,000 answer papers were personally perused by me and I found that most of the selected candidates were first rate talents. We also found many errors in the examination process. The Commission fully cooperated with the Hon'ble High Court in the legal process which was in progress.

Our effort was to make the entire process totally transparent and error free. I felt it was important to ensure that the disease of corruption was ruthlessly curbed at all costs. Most people in Manipur felt that there was no way one could get a job without resorting to bribery and corruption. In my opinion, the society at large had been deeply contaminated by this disease for a long time. It became a personal mission for me to root out this menace at least in the recruitments done by the Commission. As Albert Einstein once wrote: 'The world is a dangerous place not because of those who do evil, but because of those who look on and do nothing.'

I recall that there were hundreds of senior vacancies yet to be filled by promotions in various departments of the government. Hundreds of teachers in schools, colleges, hundreds of doctors and engineers, were not promoted and some cronies of the powers-that-be were carrying out the duties of the higher posts and authorities as on 'in charge' basis. In-charge basis appointments had the potential to perpetuate the corruption mechanism in vogue in many departments. I advised the Chief Secretary that this must end. I requested the Chief Minister and other authorities as well who agreed to attend to the issue. The Commission was able to ensure that 80 per cent of all pending promotions were cleared in record time. What came as a shock to me was the fact that many valley-based candidates from the state did not have any basic knowledge about the hills and vice versa. Many Meitei candidates failed to name the seven *Salais* (similar to clans) of their ethnic group. Most of them were not aware of the geographical location of towns, demography of the hills and so on. Many Scheduled Tribe candidates were not familiar with the Manipuri language either. Such is the emotional distance between the valley people and the hill people even today. Another observation which I must share with readers is, the overall standard of the female candidates was certainly better than that of the males in most interviews.

One major deficiency that I observed is the below par standard of most of the candidates in verbal communication and confidence levels. In interviews held after a written examination I found that there were only 15-20 per cent of candidates who had studied in the state. It speaks volumes about the lack of quality education within the state. Notwithstanding this, my confidence has increased manifold in the ability of Manipuri boys and girls to compete with others, if provided some guidance. I predict that more and more candidates from the state will clear the UPSC civil services and secure their place in the merit list in the near future. All that is required is refining their written and verbal communication skills and giving them the right platform to compete with others.

Many innovative methods to eradicate corruption were adopted. A personal e-mail letter addressed to all aspiring candidates for each and every recruitment was sent from me. The mail asked them not to pay anyone in high places or any tout for jobs. I wrote that only their own merit can get them the job, and not by paying anyone in power. I asked them not to throw away their money on bribes and that bribery will never get them the job. It was rumoured that some members of the staff who had access to the results demanded cash in return for the jobs. However, in spite of my efforts, this could not be ascertained. I am satisfied that a system was put in place in order to remove any chances of corruption.

The Framework Agreement

In 1997, a Ceasefire Agreement was signed between NSCN (IM) (National Socialist Council of Nagaland-Isak-Muivah) and the Government of India. The issue of applicability of this agreement in Manipur, Assam and Arunachal Pradesh is at the root of the lingering unrest, with the potential of it triggering large-scale violence in the North Eastern region. The NSCN (IM) clearly does not recognise the existing state boundaries, thereby implying that the agreement extends beyond Nagaland to the other states as well. In 2001, when the extension of ceasefire beyond Nagaland was put in black and white, Manipur went up in flames on 18 June 2001. The Centre has also followed a somewhat ambiguous stand by allowing the so-called 'Taken Note of' NSCN (IM) camps within Manipur. The State and Central Governments, including the Security Forces, do acknowledge the existence of the outfit's camps located in Oklong, Buning, Phungchung, within the boundaries of Manipur. The answer to whether the ceasefire agreement extends beyond Nagaland has possibly been left deliberately ambiguous. Peace prevailed for some time barring continuous extortions, levying of illegal taxes and recruitment of militant cadre in Manipur as well. In 2015, the Narendra Modi Government signed a 'Framework Agreement' with the NSCN (IM) leaders which was aimed to lead to a final settlement of the issue.

As per newspaper reports, during June 2017, the Framework Agreement signed between the NSCN (IM) and the interlocutor of the Naga peace talks Shri R.N. Ravi was nearing a final solution. There were concerns that the final agreement could seriously impact the state of Manipur. As per reports the demands of the NSCN (IM) being considered by the Centre included issues like a separate constitution for the Nagas, grant of Sixth Schedule Status to Hill Districts, separate Naga flags, separate forces for the Naga territory, and so on. In my opinion, the present setup at the Centre is unlikely to accede to any dilution of the Idea of India as one entity. Developments in the state of J&K in August 2019, removal of Articles 370 and 35A of the Constitution, point to this script. The demands of the NSCN (IM), if acceded to, have the potential to create massive controversy. It was imperative that the Manipur government keep the Centre abreast of the ground realities.

The Government of Manipur accordingly ordered a Consultative Group with myself and Shri Radhabinod Koijam, ex-CM of Manipur, as members. Many well-wishers once again tried to dissuade me from taking up the assignment. According to their logic the valley people are unlikely to accept any settlement which impacts the integrity of the state of Manipur. Many well-wishers felt that the Consultative Group could be held partially responsible in case such an accord is not agreeable to the CSOs. Our charter was to act as a link between the state and the Centre with regard to the proposed agreement. The Consultative Group was able to put across the views of the Government of Manipur to the mandarins in North Block and Sardar Patel Bhawan during 2018. The written reports of various meetings with the concerned authorities in Delhi were submitted to the Hon'ble Chief Minister from time to time.

The Naga issue is a very complex matter which requires deft handling by all stakeholders. According to the South Asia Terrorism Portal website, in the year 2018, the army destroyed three NSCN camps in Nagaland, and Changlang in Arunachal Pradesh. It was reported that an unspecified number of temporary camps were destroyed in Paren district, Nagaland on 1-2 March 2019. Two undesignated camps in Ukhrul district (Manipur) were destroyed on 25 May and 28 April in 2019. On 30 June and 5 July 2019, the army and Assam Rifles destroyed a camp each in Ukhrul and Senapati districts of Manipur. It is apparent that no shooting was involved in all these instances even though gross violations of the ceasefire agreement were carried out by the NSCN. It is apparent that both the sides were flexing muscles. NSCN leader Mr Thuingaleng Muivah made a statement in 2019 end that the talks have been reduced to the Governor level with the appointment of Shri Ravi as Governor of Nagaland. He further stated that R.N. Ravi was 'capricious and bossy'.

It is becoming more and more difficult to predict the outcome of the Naga talks. Even more difficult is to predict the consequences of the accord as and when it comes. In my doctoral thesis 'Ethnic Conflict in Manipur', I have attempted to clearly trace the root causes of ethnic divide in the state and suggest measures towards a conflict resolution. The findings of my research indicate that it is not too late to bring about Hill-Valley unity within the state of Manipur. The Naga groups such as NSCN (IM) and many of the Kuki groups were already in talks with the Central Government. However, all the important Meitei groups were against talks. They felt that no purpose would be served by talks since the stated position of the Government of India was known. At the same time they were also apprehensive about the outcome of the talks between the NSCN (IM) and the Government.

Today, the valley-based outfits have significantly reduced their activities both within and outside the country. As per reports, recruitment to these groups in the recent past has dwindled to negligible numbers. This, coupled with the fact that in recent years the Myanmar and Bangladesh governments have forced them to vacate their camps in their countries, it is unlikely that the VBIGs (Valley-Based Insurgent Groups) will be in a position to seriously challenge the writ of the Security Forces. The entire discourse in the minds of people of Manipur revolves around the possibility of the state losing its identity as they understand it. Naga talks is one of the major concerns in the minds of the people. Most Manipuris cutting across ethnic lines favour an amicable solution to the Naga issue without harming the state as an entity. Mr Radhabinod is a seasoned politician; a former Chief

Minister of Manipur, he has been the Advocate General of Nagaland in the past. Both of us worked as a team during every interaction with the State and the Central Governments.

On 5 August 2019, we had the first round of interaction with the major CSOs of the valley. AMUCO (All Manipur United Club Organisation), UCM (United Committee Manipur) and others were represented by their presidents at the meeting chaired by the Hon'ble Chief Minister of Manipur. The plan was to interact with the CSOs of the hills too so that our recommendations could be placed before the State and the Central Governments as early as possible. The challenges before the Consultative Group were many. The emotionally charged atmosphere in the state could lead to more distrust and suspicions. The Meiteis find it hard to erase their idea of the state which existed here for centuries, while the hopes and aspirations of the tribal brethren in the new order cannot be wished away either. I do hope that the authorities find an amicable and honourable solution to the issue. Some give and take approach is the only way forward. I reckon that the 'Final Accord' would likely have seen the light of the day by the time my book goes to print. The outcome of the entire process will be known only then. I know, and everyone who cares knows, that the situation in Manipur and Nagaland today is a like hornet's nest, and likely to turn even more complicated with one wrong step on the part of the Central Government.

As per available sources in the public domain, some of the important demands from among the over 20 lists of demands of the NSCN (IM) include the following:

- Integration of all Naga Inhabited areas into one single state entity
- A separate Naga constitution
- A separate flag for the Nagas

- Separate passport for the Nagas
- Joint defence of the Naga territory with Indian Security Forces
- Resettlement of their cadres in the Indian Security Forces
- A Pan-Naga Supra body that will have legal status

Most of their demands if acceded to will affect the neighbouring states in some way or the other. In fact, it will dilute in large measure the idea of 'One India' espoused by the ruling dispensation. It is also important to note that the Nagas do have reasonable rightful historical claims to protection of their identity and the land. It is said that Naga insurgency is the mother of all insurgencies in North East India. Their fight with the Indian state has been the most brutal and the longest in the region. The NSCN (IM) continues to be the most powerful group even today among the 40-odd militant groups in the region. The issue remains as to how these diametrically opposite views of groups of people is able to find a common way forward. The stand of the NSCN (IM) on one hand and that of the neighbouring states on the other is poles apart.

The views of most of the valley and Kuki CSOs are well known. Their views have been expressed in various forums including the local media from time to time. The position of most of the valley and Kuki Civil Society Organisations is directly opposite to and in contrast to the demands of the NSCN (IM). The views of the Naga CSOs are in line with that of the NSCN. It is therefore natural that the CSOs in Manipur would oppose any such demands that seriously impact their state. In the light of abolition of Article 370 in respect of the state of J&K, there has been speculation among the stakeholders as to whether the Indian Government would concede to the demands of the NSCN (IM) (especially regarding the special provisions enshrined in Article 371A relating to Nagaland, after the special provisions for J&K were scrapped). According to *The Hindu* newspaper dated 25 August 2019, quoting NSCN (IM) sources, there has been no headway in three years since the Framework Agreement was signed on 3 August 2015. The report further added that the NSCN (IM) for the first time stated that a 'Separate flag and Constitution' were necessary for an 'Honourable solution' to the 22-year-old Naga peace process. The core issue of the Nagas appears to cement the Naga identity in order to achieve their dream of creating a Nagalim (Naga Country) in some distant future.

The question arises as to what does Nationalism mean to some of the North East people like the Nagas, Manipuris, Mizos, Assamese, Khasis and so on. Is ethnic identity the only basis for Nationalism or for that matter sub-nationalism? B.R. Ambedkar opined that India was a nation of 3,000 Jatis. In the words of Winston Churchill, India was merely a 'geographical expression'. The idea of a nation state as we know since the Napoleonic days is based on common descent, language, and shared culture. The idea of self-determination today also co-exists among various groups of people on the planet. Since all the 'Nagas' in the region do not share a common descent, language and shared culture, it is believed to have been manufactured by the NSCN(IM) during the last 40-50 years at times through coercive means. Mr Muivah has successfully espoused the idea of Naga self-determination to the people. It is in this context that NSCN (IM) insists upon a separate constitution and a Naga flag besides other demands.

Manipur was among the 556 princely kingdoms which merged with the Indian Union. The 2,000-year-old Manipur kingdom which had a separate flag and constitution for many centuries lost both with the merger. Do such kingdoms revive their demand for identity, flag, etc. in today's India? What if some of the 556 kingdoms that merged with the Indian Union decide to reclaim their past 77 years after Independence? Each of the 3,000-odd ethnic groups in India is bound to consider themselves as 'unique' in their own way. In the process of a democratic path to nation-building some will gain and some will lose depending on the context. Some methods to protect the smaller communities and groups are imperative to bring about a sense of equity. The larger issue remains as to how the differing interpretations on the key words already inked in the Framework Agreement, such as 'Shared Sovereignty', 'Inclusive Arrangement' and 'Co-Existing as two Entities', will be incorporated in the final deal. While Mr Ravi's interpretation appears ostensibly quite in order, the understanding of the NSCN (IM) leadership appears to be just the opposite. Further, the Central leadership appears to have already in the past acknowledged the claim of the NSCN (IM) that they represent the Nagas. The NNPG (Naga National Political Groups, a conglomerate of seven Naga organisations) is emerging as a power centre within Nagaland. NNPG is at loggerheads with the NSCN (IM) at this point in time. NSCN often asks as to how the NNPG which were not part of the talks for the last 22 years could be a party to the talks.

The turn of events became more worrisome after NSCN (IM) issued a statement on 2 March 2020 as a counter to Mr Ravi's statement made two days earlier. Mr Ravi had said on 29 February 2020 that 'GOI cannot wait indefinitely for NSCN-IM'. 'NSCN (IM) are trying to imagine things which are not in the Framework Agreement and misleading the people', making 'new and mischievous interpretations' and that 'Delay is entirely on the part of NSCN (IM), it appears that they are not prepared for a settlement'. He also stated that, 'there is no mention of shared sovereignty in it'. In an interview to *Nagaland Post* Mr Ravi also accused NSCN (IM) of making 'new and mischievous interpretation of already agreed positions on contentious issues'. He also stated that there is 'no mention of integration of Naga Areas, No mention of Nagas as separate entity' and so on. The NSCN (IM) reacted strongly, short of putting the Framework Agreement in the public domain and claimed that the following are part of the proceedings thereof:

- The Framework Agreement recognised their 'unique history and situation' which entails that the Nagas were never part of India
- The Agreement acknowledged that there will be 'sharing of sovereignty'
- Recognition of integration of all Naga areas as the 'Legitimate rights' of the Naga People
- A Statutory Pan-Naga Supra Body has been mutually agreed upon where Legislative, Executive and Judiciary will be with the Naga Government
- The Anthem, Emblem and Insignia have already been agreed upon

(Source: *Newmai News Network*, Dimapur, 2 March 2020)

In the run-up to the date i.e. 28 October 2019 announced by Mr Ravi on 5 August 2019 to finalise the agreement, there prevailed a state of fear and anxiety among the people of both the states of Nagaland and Manipur. While the NNPG of Nagaland was ostensibly prepared to sign a pact with or without the main group NSCN (IM), the latter was seemingly determined to bargain for a Naga flag and a constitution of the Nagas. The NSCN (IM) feels that Naga identity cannot be preserved without the two.

31 October 2019: A Date Few in the States of Manipur, Nagaland, Assam and Arunachal will Forget

In the run up to the conclusion of the talks between the Government of India and the Naga groups, the Interlocutor had set the date of 31 October 2019 to conclude the talks. It had been one of the most anxious moments in the minds of the people of the region as to what could be the outcome of the talks. There were fears of a breakdown of the talks in the immediate few days preceding the date. The deadlock was due to the insistence of the NSCN (IM) for a Naga flag and a constitution.

A few days prior to the date some leaders from the outfit defected to the Nagaland-based NNPG which was negotiating separately with the interlocutor, thus leaving the NSCN (IM) leadership confined among the Tangkhuls of Manipur. The ZUF (Zeliangrong United Front) outfit, most of whose cadre were from the hills of Manipur, also expressed their support for the NNPG prior to the conclusion of the talks. There were reports that the Interlocutor and the Government of India were firm that the accord will be in place with or without the NSCN (IM), thus putting considerable pressure on Mr Muivah of the NSCN (IM). The NNPG was also ready to ink a deal without Mr Muivah. The possibilities were a worrying sign indeed.

I was in the Capital for another reason when these talks were in progress. I was requested by the Hon'ble Chief Minister of Manipur to be part of the all-party delegation from Manipur to meet the Central leadership at that time. The purpose of the delegation was to convey the concerns of the state regarding the unfolding and unpredictable scenarios in the case of either a done deal or failed talks.

In the meantime, the visibility of the security and police forces in the state had increased many fold to face any untoward situation. The COCOMI (Coordinating Committee on Manipur Integrity), a conglomerate group of the most important Civil Society Groups in Manipur, had called for a statewide vigil to protest against any agreement which could seriously impact the 'Idea of Manipur' as they saw it. The talks did conclude on 31 October. Details are not known to the Government of Manipur or to anybody else, apart from a few Central leaders. It is apparent that both sides agreed to step back a little from their respective stances. The media remained abuzz with all possibilities. There have been efforts by all the affected states to obtain the details of the agreement. But, the Centre has been assuring all stakeholders that they will be consulted before the final agreement is inked in the near future. However, the fear, anxiety and apprehension in the minds of the people continues unabated. During the period, local newspapers carried headlines such as 'Alert rallies, sit-ins staged at various places', 'AICC vows to fight for state's integrity', 'MPP (Manipur Peoples Party) warns against rushing to final deal', 'Won't Back off (from agitations), *asserts COCOMI*'. In the meanwhile, firecrackers were burst and celebrations were reported in some Naga areas on the evening of 31 October in the belief that most demands including the Naga flag had been agreed to by the Government of India. The question remained as to what the final agreement would look like. I was an invitee to a few crucial meetings of the Government of Manipur around the time where my views were also taken.

There are people holding very high posts in Delhi who genuinely believe that the Meitei Kings did not have administrative control over the Manipur hills before the British came on the scene. Nothing can be farther from the truth. Apart from the 2,000-year-old chronicle of the King, there are numerous evidences to prove that the king had his jurisdiction up to the southern bank of Brahmaputra in the north, beyond the Chindwin river in present day Myanmar to the east, and upto the western bank of Barak River in present day Cachar district of Assam at one time or the other. These facts are clearly brought out in the various treaties signed between the kings of Manipur and the British Agents as also the Pembarton Report of 1835 submitted to the Governor General of British India. The Pembarton report and James Johnstone mention about the Manipuri Police Outposts in the territory of Manipur at Jiri (Barak River), at Mao near Kohima, Kongal Thana in the North East and at Moreh in the East–which broadly define the borders of the erstwhile kingdom. It is also mentioned by Johnstone that the Manipuri king subdued the Angami Hills in the north of the kingdom, and exacted annual tribute to the king of Manipur. Yet another document presented to the British parliament by English historian Walter Hamilton in 1820, notes that the boundary of the state was upto Sylhet in the west and Arakans to the south! Therefore, to conclude that the king did not have any authority over his territory would only be a distortion of history and facts.

Heeding some wild suggestions to push for the final Naga accord anyhow to prevent breakdown of the talks, and face any outcome thereafter, would not be wise. Any decision by the Government of India based on factually distorted inputs could result in a faulty and uninformed accord. The North Eastern region and Manipur in particular is a very sensitive region and the authorities must ensure that their decisions are based on facts rather than recently constructed beliefs. A Nation State, as Gurcharan Das wrote in the *Times of India*, 28 August 2019, is an 'imagined community'. He further states that in the context of Kashmir, 'the only way to integrate is through good governance, equality of all, education and opportunities, health, and craft conditions for prosperity'. Mr Ravi, the interlocutor and the Governor of Nagaland, had allayed fears that the scrapping of Article 370 would cast a shadow on the peace process and the state. In the light of the seemingly non-negotiable stands of the other stakeholders, it appears that any solution to the vexed problem cannot satisfy all the parties concerned. All parties seem to be in a maximalist position today. At the same time, no effort should be spared to arrive at a peaceful accord by way of give and take. Simultaneously, the states involved should have no worry if their executive powers remain largely intact while a Pan-Naga advisory body is put in place for cultural matters. Delegation of additional powers to the District Councils without significantly compromising the powers of the state would also bring better administration in remote areas. The fragile relationship between all the stakeholders needs careful handling. Cultural education of the people, cross-cultural and ethnic dialogue are the need of the hour. The virtues of tolerance need to be highlighted among the masses. Dr Martin Luther King once said that God is the only solution while referring to the American racial intolerance. He also called for 'Biblical Love' as the medium through which the racial hatred would end. In the words of Martin Luther King Jr, 'Let freedom ring from every side, hills and mountainsides ....'

In his book *The Difficulty of Being Good: On the Subtle Art of Dharma,* Gurcharan Das writes, in the scene after the battle of Kurukshetra in *Mahabharata*, Yudhishtir is quoted to have told Arjuna:

> To get a piece of the earth we totally abandoned men who were equal to the earth, men who should have never been killed. And now we live with our kinsmen dead and our wealth exhausted ...Like dogs we greedily went after a piece of meat!'

There are high social and historical stakes involved here in the region and in Manipur in particular. Any negative fallout from the accord and talks in progress could have major security implications for the country. Maybe, the Corona crisis will provide more thought and find novel ways to resolve issues!

Another viewpoint

Manipur is truly a land of mysteries. Col James Johnstone in his book *My Experiences in Manipur and the Naga Hills* (1896) wrote:

> The early history of Manipur is lost in obscurity, but there can be no doubt that it has existed as an independent kingdom from a very early period.... We have no means of ascertaining what government it had before the year 700 AD, but it is believed that a monarchy prevailed at that era. About the year 1250 AD, a large Chinese force invaded the country, and was signally defeated; all who were not killed being made prisoners...

In my view, Manipur as a state commenced its downward spiral after the seven years devastation from 1819-26 till most part of the present century when the 'population of Manipur contained (was reduced to) only 2,000 inhabitants, the miserable remnants of at least 400,000-600,000, that existed before the Burmese invasion' (p. 86).

The people here have gone through multiple cycles of violence, wars and unending miseries in the last 200 years. Prior to 1819, it has been recorded that continued battles with brief intervals took place between the Avas (Myanmarese) and Meiteis. In 1738 King Garibnawaj marked the eastern door of the Kaungmadaw Pagoda near Mandalay in Sagaing in Myanmar with his sword after he defeated the King of Ava. Later in 1757, the Ava King Alaungpaya took a few thousand Manipuris as captives from Manipur. The Ava King raised a Kate (Meitei) cavalry of 2,000-strong with which he invaded Siam (Thailand) in the following year. A combined force of Kate cavalry and the Myanmarese king defeated the Chinese invasion of Sagaing in 1769. The period from the beginning of 1819, when the Ava King Bagyidaw invaded Manipur, which also heralded the downslide of the powerful kingdom of Manipur, is popularly known as the Seven Years of Devastation. According to Myanmarese chronicles of the kingdom, King Bagyidaw carried away thousands of skilled workers, carpenters, painters, goldsmiths, astrologers and indigenous doctors to Myanmar and settled them there. The Cassey cavalry of the Burmese kingdom which defeated many other kingdoms comprised entirely of Manipuri horsemen.

The treaty of Yandabo 1826 between the British and the Burmese directly impacted the geopolitical equations in the region including Assam and Manipur. The Anglo-Manipuri war of 1891 in which the Manipuri army was defeated by the British forces and the Kingdom of Manipur came under the Crown as an independent kingdom, was the final death knell. The power of the kingdom was vastly reduced and the ascent of the British power came on display. During World War I the king of Manipur provided 2,000, mostly tribesmen, as Labour Corps to the Allied Forces. The Naga movement was an outcome of the idea of nationhood of the Nagas inspired by what was experienced by the Labour Corps in the European war front. The Kuki rebellion in the region on the other hand was the outcome of resistance of the Kukis from being enrolled in the Labour Corps. The devastation and the social impact of World War II which left the region in shambles are well documented. Events leading to the merger of Manipur with the Indian Union in 1949 have left the populace divided on the ideological front. Further, the psychological impact of the 1962 war with China, tumultuous changes in the region post the 1971 war and creation of Bangladesh, and above all the unending conflict for ethnic identity and sub-nationalism resulting in the 70 years of insurgency, have all contributed to the socio-psychological situation that exists today. These events in history have left indelible scars in the polity and in the minds of the people of the state of Manipur.

The impact of the British administration of the region where at least six-seven million people were converted to Christianity over a period of 100-150 years, their 'Divide and Rule' policy through enactment of different laws for the hills and the valley, introduction of Inner Line Regulations and massive exploitation of resources of the region, have further exacerbated the already stressed social and historical tensions of the state in particular and the region in general. Post-Independence, the rules and regulations of the British perpetuated by successive Central Governments, unknowingly or knowingly, led to the increasing divide among people in the state as well as region. The idea of the state of Manipur as it existed during the 2,000 years rule of the kings of Manipur was thus diminished to a considerable extent. The exploits of King Garibnawaj against the Ava (Burmese) kingdom and whose territory extended as far as the riverbanks of Brahmaputra in the North, Surma valley and Sylhet to the west and upto the River Chindwin (Ningthi Turel) to the East, are but consigned to nothingness in history, as we mark 200 years of the Seven Years of Devastation of Manipur at the hands of the Burmese Kingdom in 1819. Many Kuki and Zeliangrong tribes are spread across the state boundaries of Assam and Nagaland. Today, ancient scriptures like 'Hidak–lon' (Language of medicine), 'Yumsa-Lon' (Way to make a house), treatise on Ways of Warfare and many other ancient texts of Manipur are extinct. There are more Naga-affiliated tribes today in Manipur than in Nagaland. Their ethnic and cultural affiliations are a reality and need to be respected.

To recover from such a level of divide and devastation of the state in terms of the socio-psychological and economic impact, requires a great amount of resilience and courage for the society as a whole. It amazed me to see the changes for the better that have taken place in my state in the last two to three decades. No other state or society would have gone through such transformations in such short a time.

I shall try and put forward my views on some issues. Manipur has seen tumultuous changes in every way. Change is the only constant in human existence. Some changes are for the better and some others are for the worse. I feel sad that society as a whole fell short of managing the changes, and the transition to modernity resulted in upheavals all around.

Manipuri diaspora in Bangladesh, Tripura, Assam, Burma and other places, who migrated to these places in wake of the various conflicts and wars, have partially adopted their new habitat as their own even though deep emotional ties with Manipur still exist. The concept of tribal, non-tribal syndrome has reached a situation where their meeting point is distant. The earlier practice of discrimination by the Vaishnavites against the hill people and others is slowly being naturally turned into somewhat of a 'reverse discrimination', as each society grasps for a domineering position. The tribal Christian brothers and sisters have largely adopted the western mode of Christian social and cultural practices with a strong blend of their original tribal customs. The Hindu Vaishnavites in the state remained mostly out of the larger Hindu fraternity. Sanamahi practices (Meitei faith based on ancestral worship) continue side by side with the 'Gaurya' (Vaishnavite) rituals. In fact it was due to this fact that in 1993 the late titular King Okendra announced that the state religion of Meiteis was 'Sanamahi' or Meitei Dharma and 'Gaurya' (Vaishnavism) was no longer the religion of the kingdom. Though the king was only a titular one and had no obligation to the people, it is assessed that today close to 13-14 per cent of the Meiteis are Christians and another 15-20 per cent of the population are following Sanamahi/Meitei faith. The number of converts to Christianity and the Meitei faith are only increasing day by day. As per the census of 2011, there were 41.39 per cent Hindus and 41.29 per cent Christians in the state. In 1951, Hindus were 60.13 per cent and Christians were 11.84 per cent. Christians had a decadal growth rate of 122.3 per cent while Hindus comprised 38.52 per cent. The impact of such a sharp shift in the society can be huge.

The religious divide is bound to impact the social cohesiveness among the Meiteis in the long run. There are many elements of congruence between 'Gaurya' faith and the 'Meitei' faith among the Meitei society and they can co-exist just as Buddhism and Shintoism in Japan in spite of counterclaims by the Revivalist agenda. There can be no other reason for such a massive shift in religion except either through proselytizing, changing preferences or increase in population. Either way, the situation can be alarming. In the census year of 2011, it was revealed that there were 35 religions/sects/beliefs/ and 'other religions and Persuasions' in the state.

Ethnic identity and sub-nationalism have dominated the political and economic discourse in the state in the last 75 years. This has resulted in identities and history being constructed through means other than facts. Ethnic clashes during the 1990s will remain a dark period in the history of Manipur. The Naga-Meitei relationship in the backdrop of NSCN (IM) demands remains a major cause of concern. Kuki-Naga clashes of 1992 have been an unfortunate event in the history of Manipur though the rivalry and clashes between the two have existed for more than two centuries. Meitei-Pangal clashes in 1992-93 have further accentuated the divide between the two. Kuki-Zomi clashes of 1997-98 caused irreparable scar in the minds of the affected groups. Mutual suspicion between the Kuki groups and the Meiteis due to lack of mutual trust cannot be wished away. Continuance of such unrests, suspicion and clashes among the ethnic groups will only bring despair to all parties. The consequences of inter-ethnic clashes, God forbid, in the future are difficult to predict.

Another flashpoint erupted between the three major ethnic groups during the proposed centenary celebrations of what some called the Anglo-Kuki war of 1917-19. During the reign of Maharaja Churachand Singh the British government requisitioned a few thousand tribals of Manipur to serve in the Labour Corps in Europe during World War I. Accordingly, 2,000 Naga tribals were sent by the Maharaja. However, the Kukis resisted the demand of the King and the British government. This led to punitive action by the British against the Kukis. The Kukis in retaliation fought the British forces during the said period, which the Kukis claim as the Anglo-Kuki war and others call it the Kuki Rebellion in Manipur. The Kuki groups led by the Kuki Inpi decided to celebrate the centenary of the Anglo-Kuki war in 2019, amid the backdrop of the charged atmosphere created by the Naga Peace talks. The Kukis also term the Anglo-Kuki conflict as one 'in defence of the Ancestral Land and freedom'. Not unexpectedly, the Tangkhul and some other Naga groups opposed the erection of memorial stones by the Kukis with the aforementioned words, as most Naga groups consider the Kukis as recent migrants to the hill areas of Manipur, as late as 1830s. *Newsmai* News Service vide their paper dated 15 October 2019 carried a press release by ANSAM (All Naga Students Association of Manipur), a student body, that said 'commemoration of the 1917 war cannot bring peace if one intends to claim what does not belong to one'.

According to the Naga groups, the entire story of the Kuki war is 'concocted and lop-sided'. On the other hand the Kukis claim to be part of the original inhabitants of the state. The state government ordered that the memorial stones with the inscription of the particular words be removed before the celebrations to avoid any untoward incident. The claims and counterclaims have some evidences to back the respective stand of the two sides. However, the undeniable historical fact is that many Chin-Kukis did enter Manipur from Burma in early 19th century with the connivance of the authorities. It is also well known that Chin-Kukis have traditionally been a wandering race. Nonetheless, it will be naïve to claim that all Kuki tribes are of recent origin in the state of Manipur. This struggle for supremacy and control of the resources and the land in the hills is being played out in recent times in every facet of the Naga-Kuki relationship.

The 2016 World Bank report said that violent ethnic conflicts have increased the world over during the period. According to Joseph Rothschild, a renowned expert on ethnic relations, issues such as ethnicity or religion are primarily exploited for political or economic rights. The ultimate aim is generally to secure political or economic rights as perceived by each of the groups. In short, to address the human needs. There is thus a need for constant engagement and dialogue between the state and the aggrieved parties to address the human needs in a more scientific and realistic manner.

More Challenges on the Horizon

In the new market system of economy, land and resources assume great significance. The fight for land and resources among the ethnic groups has assumed violent dimensions. All for money and domination of the other. Due to limited manufacturing industries in Manipur, the economy thrives on consumption of items smuggled from neighbouring Myanmar and imports from other parts of the country. Black money, underground money and corruption money are the key drivers of the market in many ways. Agricultural land in the valley is diminishing at an alarming rate since most businessmen need land for their businesses. Most well-to-do hill people, including village chiefs, politicians and businessmen have their homes in the valley.

Many lakes and waterbodies have disappeared in the valley due to illegal encroachments. The administration remains unable to address the issue of illegal and unabated encroachments in the *Khas* (government) land or in forest areas by encroachers even today. The very existence of Loktak Lake which perhaps is the lifeline of the flora, fauna and ecosystem in the valley appears to be corroding faster than previously thought. Imphal city is yet to have a proper garbage disposal and sanitation drainage system in place. Pollution levels in the valley are expected to increase manifold due to poor garbage disposal, particularly due to burning of plastics in large scale. The rivers and the water bodies have been encroached upon and some have been turned into dumping ground for garbage. Even today, I see forest fires in the reserved forest hills surrounding the valley or beyond almost on a daily basis for *Jhum* (slash and burn) cultivation.

The valley has approximately 2,000 sq.km. area. Lakes and other water bodies comprise upto 400 sq. km. The remaining 1,600 sq.km. area sustains approximately 15-18 lakh population today. Therefore, at a conservative estimate, the population density is around 1,000 per sq.km. which is unsustainable in the long term even though average population density of the state is less than 115 per sq.km. (Census 2011). Israel, with a population density of 395 per sq.km., is worried that it could be on the way to social and ecological disaster due to the density of population. As population density increases, a divided society becomes more violent and unpleasant to live in. One can predict the future of the valley of Manipur considering that the population density, including the transient resident population, might have already crossed 1,000 per sq.km. The average population density of India is 460 per sq.km. today. The density of population in the valley is comparable to the heavily populated states of Uttar Pradesh and Bihar. The social impact of such population density without a commensurate and sustainable ecosystem is well known.

Illegal trade of drugs and contraband goods has multiplied many times over along the Indo-Myanmar border in recent times. The Free Movement Regime (FMR) between India and Myanmar is misused by all and sundry for smuggling, gun running, illegal migration, human trafficking and other unlawful activities across the border. The estimated volume of illegal trade amounts to over Rs 6,000 crore a year. In my opinion, the above issues are a result of decades of mal-governance and neglect by the powers that be. One of the reasons behind the huge illegal migration to Manipur is the attractiveness of the state in terms of earning easy money through illegal means, and the ease with which these people are able to merge undetected among the local populace. It is said that one can get anything from some officials for a price. Narcotic smuggling at the border should be stopped immediately for the future of the state. Narco-terrorism cannot be allowed to take roots here.

It has been reported that there could be approximately a few thousands of refugees from Myanmar and Bangladesh settled in Manipur valley areas in the recent years. Due to the National Register of Citizens (NRC) in progress in Assam, Manipur has become a safe haven for the illegal migrants. Since other states in North East India are tribal dominated, it becomes difficult for the migrants to operate from the tribal states and areas due to the existing laws. In Manipur, the valley is the most affected area since the hill areas, which comprise about 90 per cent of the land mass of the state, do not permit non-tribals to settle down there. There is thus a real issue of sustainability of such a large population in the valley in the backdrop of the meagre resources and infrastructure in the valley. It is also felt that the train connectivity from Silchar in Assam to Imphal, which is likely to be completed in a few years' time, will also add to the pressure on the already depleted agricultural land in the valley.

The Manipur People's Protection Bill was passed on 23 July 2018, a long-standing demand of the valley people particularly the Meiteis who do not enjoy protection under the Constitution. The Bill defines 'Manipuris' and 'Non-Manipuris' and seeks to regulate entry and exit of Non Manipuris into the state. The Bill aims to enforce the Inner Line Permit System which is in vogue in the states of Nagaland, Mizoram and Arunachal Pradesh. The earlier version of the Bill was passed by the previous Congress government, but it triggered major protests by some of the tribal groups on suspicion that the Bill was designed to grab tribal land by the valley people. The Bill was sent back by the President. The 2018 Bill seeks to have 1951 as the cut-off date while demands are being made that the cut-off should be 1971 since the 1951 records are not accurate. The Inner Line Permit System was created by the British to protect their commercial interests from the marauding tribes and prevent rivals from entering into the areas where tea, oil and timber industries were located in the erstwhile Assam under the Bengal Eastern Frontier Regulations 1873. There undoubtedly exists the need to have a regulatory mechanism to protect the people of the valley as is the case in the hills.

Today, majority of the valley villages can be grouped in a similar socio-economic ecosystem as their hill brethren. Meitei culture and traditions were once considered more developed than the rest of the tribes in the neighbourhood extending hundreds of miles all around. They were more educated, had far more sportspersons and officials in the government establishments in the 20th century. Today, the gap between the common valley dwellers and the common hill people in terms of parameters like education, sports, culture, economic well-being is much narrower. This in itself is a sign of progress. The constitutional protections and privileges provided to the tribal people have ensured that tribals have surpassed the Meiteis in many benchmarks such as number of officers in the All India Services, group cohesiveness and individual or collective possession of land.

It is also true that infrastructure development in the hills has been neglected for a long time. Article 371 (C) (of the Constitution with special provisions with respect to Manipur) provides for a powerful body in the form of a Hill Areas Committee in the State Assembly. The body is empowered to ensure that the requisite funds and resources are given to the Hill Areas for developmental activities. The body is empowered to clear not only the Hill Areas budget, but can also obtain details of the expenditures from any department in the state. I often wonder as to how these august bodies have not been able to meet the aspirations of the people in the hills for so many years. If the same Committee exercises its powers I see no reason for developmental lag in the hill areas today. Besides the utter neglect by successive governments, the reality of massive looting and interference in developmental activities by some of the militant groups in the hills has also contributed in no small measure to the lack of better infrastructure in the hills.

The economic condition in many hill areas remains backward due to the geography as well as some traditions which are not in consonance with the modern ways of life, besides lack of governance in some areas.

Many Kuki villages have a population of less than 100-200 since many of them are temporary in nature. Hence, it becomes difficult to lay a road to connect these tiny villages with negligible or no population since there are other far bigger villages with more than a thousand people which are yet to be connected with good roads. A large number of villages have reportedly been found without any human settlements. It requires only five houses, real or fake, and paying taxes, to claim existence of a village. The Kuki chiefs own the village land. Therefore, no individual entrepreneurship can take place freely without the pleasure of the village chief. Availability of the valley land for habitation as well as for agricultural purposes is shrinking at a pace which could lead to mass social unrest in the future. In short, close to 65-70 per cent of the population is being supported on 8-9 per cent of land in the valley. The Manipuris' identity as a cohesive group is being perceived to be threatened with about 30-40 per cent of the Hindus among them having taken to other faiths in the last few decades.

The people of the state were assured of some constitutional protections at the time of merger with the Indian Union. Their leaders at the time of Independence did not want to be listed as Scheduled Tribes due to sheer lack of clarity and vision at that time. In the light of this, the Scheduled Tribe Demand Committee (STDC) Manipur has been demanding ST status for the Meiteis for long. The merits and the demerits of the demand are being debated in the society. The STDC delegation met me as a concerned citizen of the state to put across their viewpoints on their demand. There is a view among many that ST status will not bring better economic or social development or guarantee any protection to the land and its people in the valley. The others argue that there can be no other way except for ST status to protect the land and the Meitei people in the valley. Some of the tribal bodies in the state have vehemently opposed any move to grant ST status to the Meiteis. They fear that ST status for Meiteis will deprive them of their quota in Government jobs as well as open the door for Meitei settlements in the hills. These fears can be removed by way of dialogue and legal provisions. An important fact to keep in mind is that the British Raj referred to the Meiteis of the independent kingdom of Manipur as tribes only. The provision of ST status came into existence only after the Indian Constitution was framed. According to me, the well-developed and recognised status of the arts and culture of the Meiteis does not alter the merits or demerits of their claim for ST status. I do believe that arts and culture are one aspect, and the ethnicity and indigenous classification of a group of people are a different dimension. At the same time, I also predict that the Meiteis, if conferred Scheduled Tribe status, are bound to bag a very high ratio of positions in the Central Bureaucracy like the Meenas of Rajasthan, in a very short time.

I found the standard of government schools in the state to be simply pathetic. I was shocked to learn that Charangpat High School in which I studied till Fifth Standard could not produce a single student who had cleared Class 10 during the last three years, i.e. 2015-18. The school had its full strength of qualified teachers but somehow they were not able to teach the students in a meaningful way. Many of them were possibly involved in private coaching. A private school with hardly any infrastructure a few hundred yards away from the school was doing far better. Most parents, rich or poor, preferred to admit their children in the private school. I did find some initiative from the administration recently to improve the government schools, but it is yet to fructify on the ground. Children belonging to higher income brackets have private tutors employed at home for tuition purposes. I visited a few boarding schools on a few occasions. I found that the basic amenities were mostly lacking, and the schools appeared to be run as commercial ventures. The percentage of candidates who studied in Manipur and qualified in the state civil services tells the true story of the standard of education in the state. The only solution to the problem is to make sure that the teachers teach. Despite the teachers of government schools being most qualified and better paid most of them lack the motivation to teach, resulting in the sorry state of the schools. Transition to the Meitei Mayek (Original script of the language) from the existing Bengali script is in progress. But the full impact of the transition needs to be managed very carefully.

An academic environment needs to be built where all stakeholders participate in the process. Involvement of parents and local persons with credibility could be considered. Reluctance of many valley teachers to go to the hills is a major problem. A very senior and prominent elder of the hills confided in me that most valley people in government service will be reluctant to go to the hills and do their duties in the near future. It is worrisome indeed. The valley teachers blame the inadequate living quarters for the teachers for this state of affairs. The reason for the reluctance is far deeper. There are thousands of people employed as 'dummies/substitutes' of valley people who are supposed to be working in the hills. The arrangement of sharing a portion of the salary between the real employee and the 'Substitute' appears to be a convenient arrangement for the parties! I am told that these arrangements are carried out only with the concurrence of the militant groups such as NSCN. There is thus a need for sustained efforts from the authorities to fill the perceptual gap.

It is estimated that there are more than 1.5 lakh students from Manipur who are studying outside the state. The monetary outflow from the parents amounts to approximately Rs 150-200 crore per month to fund the student community outside the state. This amount is nearly 80 per cent of the state's monthly budget of salaries of all employees of the government. I asked one of the parents why he opted to send his child to Bengaluru. He told me that it was cheaper and the quality of education there was better; and earlier he had to spend a similar amount in Manipur for the coaching of his child. There is an urgent need to address the issue by creating a better education system in the state.

Manipur has a large number of medical facilities both private and government. The state has perhaps the best medical infrastructure and the highest number of doctors per capita in the region. Yet, the majority of people travel to places like Guwahati, Vellore, Delhi or Mumbai for treatment of serious medical cases. This, in spite of the fact that there are two Medical Colleges, besides the district and Primary Health Centres in the state. And there are over 1,300 doctors, including specialists or super specialists, employed in the government services in various disciplines and also many private doctors mostly in the valley area. Most of the consultations are done at the homes of the government/private doctors where they charge a significant amount from the patients. Everyone prefers to close their eyes and carry on with their lives. I have come across excellent doctors of national and international repute in the state who confide in me about the systemic issues which require to be fixed for a better healthcare in the state. In my view the issue is of Crisis of Character in the System. Absence of law enforcement capability, accountability and ethical issues are next. Medical tourism is one area where the state has great potential. The northern part of Myanmar lacks the infrastructure to provide good medical care to its citizens. Manipur with highly qualified doctors can contribute to the economy of the state by promoting medical tourism with improved medical infrastructure.

During Holi festival most of the government offices, colleges and universities automatically close for more than 10 days! The entire state machinery appears to go into hibernation during the period. Even newspapers stop circulation for at least six days during the Holi festival. Everybody seems to accept the status quo. Such traditions need an early relook and the society needs to adjust to the modern realities and the aspirations of the youth. The frequent bandhs and blockades have greater psychological impact on society besides on its economic well-being. The local political leaders and the administrative machinery have to bear some responsibilities. The law and order machinery needs to rise to the occasion if the state needs to progress. The root cause of these blockades needs to be addressed as well.

The trust of the people in the dispensation of justice, resources and opportunities is the bedrock of establishing confidence among the people. In short, good governance is the key to many of our ills. I pray that someday, the money lenders of the state will only charge legitimate and lower interest rates (they charge close to 36 per cent interest as per current rates as compared to bank rates of less than 8 per cent) from the poorer sections who pawn their small land holdings to the rich. I also pray that the merchants will pay legitimate taxes to the state and not act against the people in cahoots with some officials. And I hope that only genuine goods enter the market and not duplicate spare parts and material from outside.

Manipuris are superstitious by nature. Such practices have been well documented by the British during their presence in the state in the 19th century. The *Pakhangba,* snake deity, is never moved in westwards direction because doing so would imply retreating from the traditional enemy–Burma! Sir Geoffrey Evans in his book *Imphal: A Flower on Lofty Heights (1963)* narrates the plan of one of the British Political Agents to build a new Residency on a small hill outside Imphal; but he was obliged to abandon the plan because the Durbar objected on the ground that the hill was haunted by an evil spirit inimical to the royal family. There are innumerable instances of superstitious culture in Manipuri society.

I do hope to see a more scientific temper and truly evolved society in the state while ensuring the preservation of the identity and culture of the people. I also hope to see reduced arrogance, greed, and laziness among our people. Let our future generations unshackle the notion, if any, of feeling victimised by our looks and origin. We are unique too as much as others are!

The Opportunities

There are far more positive aspects and opportunities in Manipur which merit greater appreciation of the society. Individual and collective consciousness regarding social and economic rights and awareness are emerging as a force to reckon with. I discovered during my tenure as the Chairman of the Commission that the youth of the state are qualitatively superior to their counterparts in some other parts of the country. The superior level of their calibre would have been unthinkable in Manipur two decades ago. I deeply admire their innovativeness and the will to achieve their goals. In 2018, for the first time in the history of the state, there were 1,100 doctors who applied for jobs in the Government against the 340 vacancies available. Many of the doctors who did not get selected were found to be working in private enterprises. During the interview stage of selections, I was impressed by their merit, their intelligence, and competitive spirit. The success story of the youth of the state goes beyond the sports, arts and cultural fields. The state has the highest number of officers per capita in the defence forces where there exists no reservations or quota of any kind. I am convinced that, given the right opportunity, the youth of the state can excel far more at the national and international levels. They are physically fit, mentally strong, and adaptable to any environment, can withstand pressure of any kind, are dependable, honest, sincere and are mostly well educated. Most of the young people can cook, sing, dance, play any game, do gardening, are innovative and appreciative of nature.

Col James Johnstone, in his book *Manipur and Naga Hills*, writes about the Manipuris:

> A fairly homogenous people of great activity and energy, with much of the Japanese aptitude for acquiring new arts. They are always cheerful even in a long march, and are good humoured under any difficulties and never conscious of fatigue. (p. 98).

The number of coaching classes mushrooming in the Greater Imphal area speaks about the aspirations of the students. All they need is some right guidance, the right platform and opportunities. Given the fact that there are very few job opportunities in the state, most of them will have to find their pastures well away from their home state. As we know, close to half a million young people of the state are studying or already working in places like Delhi, Bengaluru or other metros. It is inevitable for cross migrations to take place as society progresses. I am confident that in the near future they will find much better opportunities than what they are engaged in at this point in time. In this context, it is imperative that schools, colleges, universities gear themselves up for the lead role they are expected to provide to the youth. The education ecosystem is evolving, but at a snail's pace.

The mindset of the elderly will need to be directed towards acceptance of the viewpoint of others as well. The elderly enjoy a very special place in society. They shall be expected to wield greater influence by setting a personal example among the youth in helping to eradicate the ills of society such as corruption. They need to bring in the culture of discipline, punctuality, keeping to one's words, respect and care for the elders and other things. The administrators and the scholars will need to take the perpetrators of the ills of society head-on. A narrative will have to be built up of changing society for the better. Civil society, children and the media can play a significant role through campaigns against the ills plaguing society just as children did nationwide against rising pollution.

The government machinery will need to be more responsive and transparent if they truly desire a better society. Many in Lutyens Delhi believe that Manipur is a failed state. While love of my state prompts me to disagree with this viewpoint, one cannot deny and hide such observations. The tag of 'failed Governance' does not convey that the people of the state as a whole have failed. It implies that the governance and pillars of the state have been found wanting in many ways since long. At the individual level the people have succeeded despite the state. The question now is what do we do? Is it too late? I think time is on our side. One aspect that needs to be fixed is the Governance issue at the macro level, and changes in the mindset of the people to bring in a sense of discipline in their day-to-day life. I would like to see more people appreciating the achievements of their brethren who make good changes in the society.

The law enforcement agencies will need to be more motivated and better trained to do their work as per the law. Manipur has far lower conviction rates of crimes as compared to the national average. We need to reflect on the systemic ills of such nature. The elected people have to be accountable. Business as usual will be challenged by the youth of today. The leaders of society need to set the pace. The challenges of transition from tribalism, from rural economy to modern, from pre-set mindsets of the past to inclusive and modern democratic values, will be enormous indeed. Narrow ethnic and identity issues can only bring more harm to the society. My dear friend the late Dr M. Horam had commented in his book *Social and Cultural life of the Nagas:*

> History bears testimony to the fact that periods of transition anywhere and in any sphere are delicate periods, and need watching on the part of the reformers or else there is considerable danger of the group being changed or reformed to deteriorate or even become totally degenerate.

It is my considered opinion that we still have the opportunity to manage change, if only those responsible understand the dynamics of the situation.

In the aftermath of my retirement, I had the honour of being invited to many functions in colleges, schools, universities, local clubs, civil society organisations of various hues from time to time. These interactions gave me more and more insights into Manipuri society at large. All of them had only one theme, progress of the society. Only the method of achieving their aims varies. It also depends on what is their idea of the nation and the state, their perception of identity, progress in society. The need to contextualize issues is paramount. There is a saying in Manipur that there are always two ways of doing things, two dates for every cultural event in the state, two opinions about anything, more than two faiths or religion within the small Meitei society. The society is pitiably divided almost in every sphere of existence, and this is particularly pronounced in the valley. We need to reverse this by adopting a culture of appreciating and acknowledging the views of others too. We could do well by having informed debates and avoiding undue criticism of others.

Many times it has been observed that the narratives in society are seldom backed by scientific and universally accepted logical reasoning. During an interaction programme at a book fair in November 2017, a remark which I made at the function came for sharp criticism. I had said that many teachers in schools and colleges in the state do not read adequately enough after they take up teaching jobs. Thus the students lose out on modern knowledge and latest ideas. Some people felt that my remark was an insult to the teaching fraternity. I did not retract my words. Many professors supported my views. The fact is that my comments opened up food for thought among some of the teaching fraternity.

I have always insisted on punctuality and fixed timings to commence and close any event. I found some events commenced on time but had no fixed ending time. Most of the events never commenced on time. On one occasion a high ranking person did not turn up even after 30 minutes of the given time. I was the Guest of Honour at the same event, and I left the venue and asked them to convey to the person that I had left. It was just that it had become a disease in the society, and very few expected people to be punctual on most occasions. Observance of punctuality in daily life was more of an exception rather than the rule. Few take the idea of punctuality seriously and fewer believe in the discipline to practice time management. Conversations continue endlessly in a roundabout manner, seldom sticking to the issue at hand. I very much wish that the new generation of today will soon change the scene and be more conscious of time management in their lives. I look forward to the teachers and the government officials to set the pace in this regard.

One of the most striking developments in Manipuri society has been in the religious preferences and inclinations of the common people. The Sanamahi or the Meitei cult and Christianity are beginning to gain ground among them. The Sanamahi cult is based on their faith in ancestral worship. The cult is in reality an inseparable part of Meitei society even as they follow the Vaishnavite culture as well. Sanamahi has a special place in every house of the Meiteis and some other tribes of Manipur. 'Apokpa' (ancestral) worship exists alongside the 'Gaurya' or Vaishnavite culture in the society. I have witnessed a phenomenal rise in the awareness of and in the stricter practice of the ancestral way of worship in recent years. The identity of the Meiteis is closely associated with the Sanamahi as well as the Vaishnavite faith. More and more ceremonies are being conducted in Manipuri language unlike in the past when the same were being conducted in Sanskrit or Bengali. Maybe a time will come when the best of fusion of Sanamahi and Gaurya faith will dominate the religious scenario in the society. In the contemporary world, there is a need to learn, emulate and assimilate the best ideas from others while preserving the best in ourselves.

Another remarkable achievement I noticed is in the field of individual and collective entrepreneurship to earn good income and good living through hard work. These enterprises are in many diverse fields such as education, businesses, healthcare and other service sectors. Homestay for tourists in Imphal was unthinkable a few years back. It is a success story now. Taxis/cabs on the pattern of Ola are seen on the streets of Imphal now. The attitude among the youth is changing for the better. The boys are no longer averse to working as labourers in any field. Lack of opportunities for employment in the Government sector no longer prevents the youth from taking up small start-ups. Young people who have relocated back to Manipur from outside the state have helped in the progress of such activities. To ensure that the momentum is maintained, there is a need to ensure that these activities follow the law and established regulations. There will be a need for the concerned officials to be more transparent and act as facilitators.

I visited a small workshop of a relative of mine recently. The workshop was 300 sq.ft. area where he had improvised to make sophisticated machines from the scrap parts of motorcycles and old cars. He was in the process of making a robot for differently-abled children to help them walk. Such initiatives are extraordinary and can be seen in greater numbers since the past few years. Decent financial support from the authorities without harassing the loan seekers will go a long way in sustaining such private initiatives. On the other hand the likes of Salai Holdings episode which broke out in the recent past where a few hundred crores of rupees was allegedly swindled from investors by some touts also needs to be monitored by the authorities.

During my visits to Senapati and Churachandpur towns in the hill areas, I was pleased to see the economic life and commercial activities of the people there. A very high percentage of the Churachandpur populace have jobs, local or otherwise, and there is significant disposable income in most households. A considerable number of youth from the area were found to be working abroad. The standard of education in Mao Maram area was also impressive. I was particularly pleased to learn that there are more than 13 doctors from the village of Oinam in Senapati district which was the scene of 'Operation BlueBird' during my stint in Assam Rifles in 1987. During that time there were hardly any graduate villagers leave alone doctors and engineers. I wish to see improved academic efforts in other hill areas like Ukhrul, Chandel and Tamenglong.

In my view, the failing is mostly on the part of the leadership of the society and the polity through the ages. Many important functionaries failed to see beyond their own interests or beyond themselves. They may have taken decisions based on what they thought was the best in their time but some of the decisions retarded the progress of the people at large. Failure to save Loktak Lake and many other wetlands even today is one such lost opportunity to my mind. Imphal almost forgot that the hills existed. The series of events over the last 150 years leading to Hill-Valley fault lines is another major mistake of the past. Identity crisis among the Meiteis in terms of religion/faith, the issue of OBC and demand for ST status are some other issues. History will not pardon the leadership over a period of time in the state for having miserably failed to prevent the emergence of drugs and AIDS in the state in the early '80s. Failure to preserve the flora and fauna in the state will someday adversely impact the very existence of our society in the not too distant future. Contaminated food items grown in the state as well as imported from outside are having a deleterious effect on the health of the people. Failure to control the same will cost the future generations immensely.

Corruption culture is perhaps the one most important factor which has taken a toll in every sphere of the lives of our people. I don't see things improving in other arenas till this menace is controlled. Many schemes and projects have failed to take off or see the light of day in the manner for which these were planned due to failure of proper planning and execution. Hence, the remedy lies squarely at the door of the administration also. Good governance through the medium of e-governance, visionary leadership and a few good men are all that is required to root out the menace.

Where in the world can one find woman power at play as it is in Manipur? A few years back, Manipuri women comprised the entire Indian Women's Football team. Manipuri girls play polo with little training and play exhibition polo matches across the country. Football matches are organised even among the married women during the Holi festivals. More women qualify than men in the recruitments conducted by the Public Service Commission in the state for higher posts in the government such as doctors and teachers. Manipuri women power can play a significant role in bringing about peace and progress in the state. Many of the fighters of insurgent groups are yet to reconcile to the fast emerging global and national scenarios. The leaders of these groups were among the finest minds in their own way. The insurgent groups were once a very attractive proposition among the brave youth of the state. We need to look to the future, analyse in a scientific manner the evolving world around us, and make choices for the future good of coming generations.

I have attempted to put across some of my views as to how I would like our people to progress. I have also expressed my deep appreciation to those who have struggled against many odds to bring about positive changes in the state and society at large. I have full faith and hope that sooner than later, the incredible talent and huge opportunities that are unfolding here will flower into a great future of the people. As I observe the other parts of the region, it appears to me that similar issues plague the other North East states as well in some measure. The people of the region and the Manipuris will rise together to their full capability to make our country strong and our forefathers proud of us.

Many thousand flowers are waiting to bloom. The remoteness of the region cannot be allowed to undermine its huge potential. The destiny of India is closely linked to that of the North East region. I am confident that people will learn from history and take the opportunity to build a new hope for the future generations. Oscar Wilde once said, 'I dislike modern memoirs. They are generally written by people who have either entirely lost their memories or have never done anything worth remembering!' Mine is a work of passion by a village boy, the work reflects innumerable trials of the self, observed and dissected from my own experiences. To all my young friends, I say: **Aim for the Mountains!**

Postscript

(Along with Jaideep Saikia)

As we delve into the future of North East India and mull over ways for the road ahead, one finds that similar issues exist in the entire region as is the case in Manipur. Degrees, however, differ. Every state has unique challenges. But, challenges showcase opportunities and common threads and interests. The North East has accepted the Idea of India. But the ripples of the nation building exercise persist. Ethnic conflicts, unrest in the populace due to real or imagined insecurities, and other threats to the national security paradigm remain largely active. The good news is that majority of the youth in the region want progress, employment, education and to be part of India's success story. But, addressing the pulse and aspirations of the people by the state has been far from robust.

Even as the Peace of Westphalia concluded in 1648 (Spanish-Dutch treaty that ended the Thirty Years' War) determined the principle of territorial integrity of states and attendant aspects relating to non-interference in each other's domestic affairs, later day academic debate left ambiguous the fine distinction that characterises borders and frontiers, the latter denoting a 'state of mind' rather than a physically demarcated boundary. The colonial rulers of India—in perhaps both their haste and (if commentators such as the late Neville Maxwell are to be believed!) deception—added to the predicament, and much of the 'enchanted frontiers' that makes up the North East of India continues to grapple with concepts such as borders, boundaries and frontiers.

National security in the context of the North East cannot confine itself to only the myriad insurgencies that dot its variegated landscape. It must take into account the strategic encirclement that the region is heir to. Surrounded by China, Myanmar, Bangladesh, Bhutan and Nepal (the erstwhile Himalayan kingdom is placed a little afield, near the Siliguri corridor), 88 per cent of the North East's boundaries are international, with only 12 per cent connecting it to mainland India. The borders in the North East must, therefore, lend itself to robust management. Unfortunately this has not been the case, and the porosity of the borders has been a source of considerable concern for the region, subverting thereby the national security of India, of which the North East is an inalienable part of. Whereas drugs, arms and insurgents enter the North East via the land border from across Myanmar, the border with Bangladesh—in addition to the above three aspects—also provides the route for illegal ingress of migrants into India. The demographic invasion has critically altered the population patterns of Tripura and Assam in a manner that it cannot ever be corrected. Despite the fact that the infamous IM (DT) (Illegal Migrants (Determination by Tribunal) Act was scrapped by the Supreme Court of India, there is little or no let-up in the influx into Assam and other parts of the North East from the erstwhile East Pakistan. Even the process of updating the National Register of Citizens (NRC) in Assam that 'successfully' concluded in the state has not brought about quietude to the populace, and the introduction of the Citizenship (Amendment) Act, 2019 has only ushered in further suspicion. The governments in the North East are now trying to bring about a consensus among various groups of the state on the issue of the NRC.

The threat that stems from the approach in which NSCN(IM) and their proxies are charting and carrying out a war of attrition in Arunachal Pradesh, Assam and Manipur has to be understood in the acceptable point of view. The geopolitical engineering that is being carried out in Manipur is in ample evidence. The time-tested NSCN (IM) approach has always been to hold the state to ransom by putting on an economic obstruction, and blocking it from the rest of India. In Arunachal Pradesh, it has engaged the populace with 'Operation Salvation', a clear strategy to seize territory by converting the animistic tribes in the districts of Tirap and Changlang into Christianity. The politics of the two districts are also so weighed down by the Nagas that the seismicity was felt in far-off Itanagar.

In Assam it courted the DHD (J) (Dima Halam Daogah outfit) after having signed a 'Hebron Agreement'. Cessation of hostilities with DHD (J) had in no way lessened either the alliance with the Naga outfit, or brought peace to the Dima Hasao district. Furthermore, NSCN (IM) continues to elbow into other areas by entering into alliances with other smaller insurgent groups in order to extend its area of influence. It had in the past set up umbrella organisations with an avowed objective of starting a unified move against the Indian state. The United Liberation Front of Seven Sisters, a front that included ULFA, DHD, UPDS and PLA of Manipur was one such attempt. Yet another was when the Self Defence United Front of South East Himalayan Region was formed to coordinate the activities of groups such as NDFB, NLFT and KYKL. The June 2015 attack on the Indian Army's 6 Dogras in Chandel district of Manipur killing 18 army men by a motley group headed by the NSCN (Khaplang) is yet another instance of insurgent conglomeration. The prolongation of the ceasefire—although it may be considered to be a 'wearing down' strategy by some—allowed the formation of a 'coalition of the willing' and made insurgent groups in the region more belligerent.

The woes of the North East, however, cannot be confined to only illegal migration and insurgency. Water—one of the most important natural resource on earth—has taken centre stage in Assam and is steadily becoming a weapon of politics, even as Assam readied for the Covid-19 pandemic in early 2020. It began as a mass movement engineered by the Krishak Mukti Sangram Samiti (a peasant organisation based in Assam), and its demand is to halt all mega dam projects in the North East. While some of the concerns are bonafide, the fact of the matter is that the rich blue resource of the region—awash as it is with the white waters from the Tibetan plateau—would not only be a power reservoir for the North East, but also for the rest of the country.

The issue came into focus with the apprehension that China is undertaking a mammoth project to build five dams in the Shannan Prefecture (Lhoka) in the Tibet Autonomous Region. The precise locations according to reports were to be Zangmu, Zhongda, Langzhen, Gyatsa and Jieuxu. The exercise is slated to divert the waters of the Brahmaputra (Yarlung-Tsangpo in Tibet) to the People's Republic of China's water scarce north-western and northern provinces. There is great apprehension that the diversion of the waters would affect the downstream in North East India. China has denied that it would be so, stating that it would not reduce downstream water flow.

The aspects that ail the North East are, therefore, deep and dark. These have to be addressed by way of a national policy. As a matter of fact, the authors of the postscript had recommended that a North East Security Council be set up to look into matters that are specific to the region.

In sum, the theme of this book would be incomplete if—in togetherness—one does not address the issue of the human potential of the North East. Home to entities that outline the robust profile of the borderlands of India, the youth of the 'enchanted frontiers'—it is hoped—would take a frontline step in goose-stepping the nation building exercise that India is still groping to embark upon. Spartan and marked with clarity of thought and vision, the hour of darkness should premise itself as a window of light. Indeed, this is the time to re-integrate the North East with the rest of India and free it from the shackles of being an 'outland'. But, hands do not clap in isolation. They have to proclaim one another and in unison. This is what the North East expects from the rest of India.

India of the present faces an unseen enemy: proximal yet distant—one whose crippling power is yet to play out. This has become clear after the outbreak of the Covid-19 virus, unleashed onto a blameless milieu by a malevolent force. The tenacity and perseverance of the new generation would stand both to test and validation in the next decade, even as the severity and outcome of the novel scourge unfolds itself. The North East and its human capital–in all earnestness—would stand out as the veritable 'first line of defence' and in all fairness the nation should be grateful for having such a contour that can first countenance the venom before it can enter the country's innards.

What lies ahead for the region will largely depend on some important factors such as the progress of India's Act East Policy, outcome of the talks and peace accords with militant groups and the manner in which the region's economy merges itself with not only the rest of India but with the financial giants of Southeast Asia. The issue of illegal migrants, border management, and externally inspired subversive activities will also play a significant part. The key driver is 'Good Governance'.

North East is the final frontier of India. It stands as the sentinel for the nation's security. It would be incorrect to view the expanse as a defence zone or a buffer between competing 'Asian Giants'. After all, even a province such as Manipur took on the attire that characterised far-off Vrindavan, abdicating its age-old coat. It would, therefore, be worth the while for mainstream India to comprehend such realities and not—especially in time of crises—embark upon racial profiling exercises that are being reported from regions in the mainstream. The Indian consciousness which has housed divinities such as Mahatma Gandhi should rejoice in the fact that it encompasses a populace who holds a realm close to its heart and terms it as its own. Let there be no 'us and them'. Joined in sorrow (and tomorrow, celebration) we stand against all enemies, foreign and domestic.

Jai Hind!

Acknowledgements

My memoir will be incomplete unless I thank all those who encouraged me throughout, and helped me in so many ways in my life's journey.

I thank all my elders, friends and colleagues from the military and the civilian fields who expressed their wish to see a memoir from me.

My special thanks to my brothers Nongyai, Santosh and Kiran, and my sister Anita and their families, for their love and constant support.

I am grateful to Shri Jaideep Saikia, who readily agreed to write the Postscript along with me and provided constructive suggestions during writing of the book.

I would like to thank Maj. Gen. A.K.S. Sengar (Retd) who motivated me to 'write something' after I hung up the uniform in 2017.

I am grateful to Mr Kunal Verma, who gladly provided the background photos, and Mr Robin Wangkhem, for taking my photograph for the book cover.

I thank Mr Ajit Doval, Gen. V.K. Singh, Gen. Bipin Rawat, Mr Ranjan Gogoi, Mr N.N. Vohra, and Mr Maroof Raza who kindly wrote the blurbs for my book.

The untiring efforts of Team Konark—Mr K.P.R. Nair and his team Ms Ranjana Narayan and Ms Jiza Joy—have been praiseworthy.

And above all, I owe it to my wife Dr Mangala and both my daughters, Poornima and Priyanka, for their encouragement and support in every way.

Index